Discovering Pompeii: From Effects to Causes—From Surveying to the Reconstructions of Ballistae and Scorpiones

Discovering Pompeii: From Effects to Causes—From Surveying to the Reconstructions of Ballistae and Scorpiones

Volume Editor

Adriana Rossi

MDPI

Basel • Beijing • Wuhan • Barcelona • Belgrade • Novi Sad • Cluj • Manchester

Volume Editor
Adriana Rossi
Department of Engineering
Università degli Studi della
Campania Luigi Vanvitelli
Aversa (CE)
Italy

Editorial Office
MDPI AG
Grosspeteranlage 5
4052 Basel, Switzerland

This is a reprint of the Proceedings, published open access by the journal *Engineering Proceedings* (ISSN 2673-4591), freely accessible at: https://www.mdpi.com/2673-4591/96/1.

For citation purposes, cite each article independently as indicated on the article page online and as indicated below:

Lastname, A.A.; Lastname, B.B. Article Title. *Journal Name* **Year**, *Volume Number*, Page Range.

ISBN 978-3-7258-4639-9 (Hbk)
ISBN 978-3-7258-4640-5 (PDF)
https://doi.org/10.3390/books978-3-7258-4640-5

Cover image courtesy of Adriana Rossi

Contents

About the Editor

Adriana Rossi

Since 2016, Adriana Rossi has been a full professor at the Department of Engineering, University of Campania Luigi Vanvitelli (Italy). She graduated with honours in Architecture in 1984 and completed doctoral studies and post-doctoral fellowships in "Surveying and Representation of Natural and Built Environments". She has been Deputy Director of the DicDEA, a member of the Council of the School of Engineering and Basic Sciences, a member of the departmental "Junta" of the Faculty of Engineering and, previously, of the Faculty of Architecture. She has held various academic leadership and representative roles on behalf of the university.

She is the managing director of the academic spin-off company AXIS Strutture s.r.l., established following the award for research activity with industrial impact (V:ALERE2020, DR. no. 596/2020). It has patented inventions and holds a European PCT. She is responsible for the SCORPi'o-NIDI project as the principal investigator at the national level (PRIN22). She teaches courses on subjects such as graphic sciences, digital sensing, and parametric modelling to support workflows for the analysis and design of architectural and cultural heritage, within SSD CEAR10 - Drawing.

Her publications—more than 240 to date—deal with the scientific fundamentals of representation, the geometric interpretation of formal structures, the evolution of codes and methods of representation in the light of polemological science applied to archaeological remains, digital restoration, museum design, and the dissemination and enhancement of cultural heritage.

Preface

For some contemporaries, it is incredible that "Travertine," better known as Sarno limestone, together with Gray Tuff, could be moulded like butter under artillery fire. Yet Amedeo Maiuri, director of the excavations at Pompeii between 1924 and 1961, together with other collaborators and scholars, did not hesitate to identify the cavities found along the northern walls of the city as evidence of the blows inflicted by Sulla's ballistae.

It is essential to recognize and investigate the value of ballistic evidence to promote experimental archaeology that takes historical and military knowledge into account. This is a fundamental step in ensuring an adequate and informative museological narrative. Raising a framework of meanings to bring present expectations into dialogue with knowledge of the past is a contemporary necessity, as established by the 'London Charter' (2009) and the Seville Principles (2012), now shared by scientific communities that have long insisted on the advanced digitization of inherited heritage.

Presenting the sources, illustrating the state of the art, and discussing the orientation of readings and interpretations is the practice that makes the work of analysis and inductive verification of results transparent. Imagining synergies that, through distinct paths, converge on the same objective is useful for organizing a model that digital language makes multifunctional. In the hope that favourable conditions for cooperation with the Archaeological Park of Pompeii will arise, SCORpioNIDI is moving toward the objectives funded in Italy by the Ministry of University and Scientific Research on a competitive basis.

The seminar day, which this volume commemorates, was organized by inviting expert speakers and university-level listeners with the aim of sharing the results obtained to date and promoting debate on those in progress. I would like to take this opportunity to express my gratitude to my colleagues who shared their passion for the subject, making their multidisciplinary skills and knowledge available.

I would like to express my gratitude to Flavio Russo, historical consultant to the Army General Staff, who in the new millennium has brought to the attention of his contemporaries the value of ballistic impacts, focusing mainly on the analysis of the spheroidal ones found along the northern wall of Pompeii. On numerous professional and educational occasions, he encouraged me to continue my investigations, considering my specific knowledge and skills to be decisive for the development of the research. Survey tools and methods, which are anything but passive techniques, have in fact proven to be intellectual levers. The data acquired and critically reorganized in the representative space have guided communicative syntheses of objective data expressing contemporary expectations. Digital transcripts prove to be accelerators of traditional processes as they transform observers into actors in personalized and participatory paths.

Juan Antonio García Esparza – Universitat Jaume I, Spain
Zahra Gharineiat – University of Southern Queensland, Australia
Francisco Juan Vidal – Universitat Politècnica de València, Spain
Giovanni Loreto – Kennesaw State University, GA, United States of America
Qing Mei – Tangji University Shanghai, China
Roberto Mingucci – Alma Mater Studiorum-Università di Bologna, Italy
Adalberto Ottati – Universidad Pablo de Olavide, Spain
Alberto Sdegno – Università degli Studi di Udine, Italy
Lluís Villanueva Bartrina – Universitat Politècnica de Catalunya, Spain

Organising Committee
Adriana Rossi – Università degli Studi della Campania Luigi Vanvitelli, Italy
Sara Gonizzi Barsanti – Università degli Studi della Campania Luigi Vanvitelli, Italy
Silvia Bertacchi – Università degli Studi della Campania Luigi Vanvitelli, Italy

Adriana Rossi
Volume Editor

engineering proceedings MDPI

Editorial

Pompeii: From the Survey of Ballistic Impacts Towards the Reconstructions of Roman Artillery (1st Century BC)

Adriana Rossi

Department of Engineering, Università degli Studi della Campania Luigi Vanvitelli, Via Roma 29, 81031 Aversa, Italy; adriana.rossi@unicampania.it

Keywords: photogrammetric survey; modeling and inverse analysis; digital archaeology; elastic torsion motors; Polybolo-Scorpions-Ballistae; interactive visualization

1. Conference Overview

This volume brings together the reflections of those who have been committed to building a dialogue around the results achieved (and currently underway) by the research project "Comparative Analysis and Certified Reconstructions for a correct experimental archeology: Roman Scorpions and Ballistae for the Imperial mechanical culture, origin of European identity. Governance policy for the development and sustainable fruition of Cultural Heritag", which received top marks from an international evaluation panel (Erch SH5, Annex A. D.D. no. 1012/2023) at the conclusion of the first year of research activities, funded in Italy by the Ministry of University and Scientific Research through dedicated resources from the National Program of Projects of Significant Interest (MUR, D.D. no. 104/2022) (Figure 1).

The Department of Engineering at the Università della Campania *Luigi Vanvitelli* ensured the scientific coordination of the research activities. This unit works in collaboration with the Department of Mechanical Engineering of the Politecnico di Milano and the Department of Computer Science, Science and Engineering, Alma Mater Studiorum—Università di Bologna. A joint site inspection at the research site served as an opportunity for discussion, preceded by a presentation of the preliminary results.

On 27 February 2025, in the Aula Magna of the Real Casa dell'Annunziata (Aversa, Italy), professors and researchers collectively discussed their respective progress. The perspectives of experts external to the project's research groups and professionals attending the conference enriched the exchange of knowledge, with a focus on technology transfer and the dissemination of results beyond the academic context.

The aim of SCORPiò-NIDI, the shortened name of the project, is to become a tourist attraction for the already renowned Archaeological Park of Pompeii in the province of Naples. The aim is to provide a renewed image of the city founded by the Oscans as a result of the experimental activity and the historical/military study carried out on the northern section of the fortified walls. The city walls, buried by the eruption of Vesuvius in 79 AD—specifically in the relevant section—was only unearthed during the first decade of the 20th century. The study performed focused on the possibility of detecting some of the anthropic indentations that Amedeo Maiuri (1886–1963) and Albert William Van Buren (1878–1968) attributed to the impacts of darts and stones launched by the Sullan artillery during the siege of 89 BC.

In line with the national and international objectives of the strategic plans and guidelines of European research, the digital survey of the cavities selected as the case study allowed for the generation of high-resolution, potentially printable models, suitable for documenting and disseminating what today appears to be of exceptional value: a sort of relic, being the only confirmed evidence of the lethality of existing elastic torsion weapons. It is a responsibility of the contemporary era to bring back what was deliberately forgotten at the time of its discovery to public attention. The widespread rejection of the horrors produced by the world wars understandably led to the concealment of the lethality of deadly Hellenistic weapons.

Figure 1. Conference poster.

Today, shared and interoperable workflows allow for the integration not only of models but also of fragmented knowledge. Therefore, it is increasingly necessary to outline a general framework that offers a comprehensive and detailed overview to visitors of the Archaeological Park of Pompeii. Recognizing the value of ballistic evidence derived from experimental activity and historical/military research is a fundamental step toward an appropriate, objective, and reliable museological narrative. The renovation of pedestrian and cycling routes for tourists around the northern section of the city walls is an ongoing effort; illustrating and making ballistic evidence intelligible contributes to defining a cultural identity that, although distant in time, still has much to reveal. Digital surveying makes it possible to transcribe evidence of exceptional value into updatable documents. Digital technologies allow for an in-depth collection of detailed features and the use of multipurpose and multiscale models. Among these, a key element for investigating the effects of stone balls and dart launchers is the 3D acquisition of high-resolution casts, used for reverse modeling. The study of geometric configuration is essential for solving reverse problems—specifically, in our case, to define mechanical parameters through which to find the characteristics of terminal ballistics.

The originality of our research lies in what appears to be a unique possibility of calculating the volume of fractured material in order to estimate the kinetic energy required to cause breakage. This static–mechanical characterization, inductively validated, helps us confirm the calculated values by drawing upon Hellenistic science formalized on experimental foundations.

The results will provide objective elements, as they are based on verifiable effects. A pilot series of reliable and functional reconstructions may be appropriately patented to serve as a reference for the "certified" production of museum installations within and beyond Europe. In parallel, virtual scorpions and ballistae will support dedicated platforms, both online and offline, and physical demonstrators for an engaging and interactive experience on site, near the fortified walls.

Accurate documentation aims to preserve, for future memory, the "relics" that have miraculously survived Roman restoration, the volcanic catastrophe, World War II bombings, the effects of weather and natural elements, and, not least, interventions that were not always aware or respectful of the unique nature of the remains within the city wall section between Vesuvio Gate and Ercolano Gate. Moreover, human recklessness and natural disasters could erase—this time forever—the evidence of a refined science of which we possess only a few damaged archeological remains.

At the end of this conference, both the limitations and potential of digital acquisition and processing methods were highlighted, drawing attention to the challenges encountered in various case studies surveyed in Pompeii when analyzing irregularly shaped indentations and stone blocks of varying sizes, materials, and densities. At the heart of the debate were heuristic interpretations of the data obtained from the survey. Among them, one particularly intriguing possibility emerged: the potential identification of impact traces left by a repeating catapult, known in ancient sources as the *polybolos*.

This weapon, attributed to the ingenuity of Dionysius of Alexandria (an engineer active in the arsenal of Rhodes), has been described in detail as the first known automatic missile launcher in history (Philo of Byzantium, *Belopoeica*, 73.34). The "multi-launcher" differed from traditional elastic torsion catapults in a fundamental way: it could launch multiple darts in rapid succession without the need for manual reloading. It thus represents the peak of ancient science, capable of conceiving a sophisticated semi-automatic system that may have been known to the Roman military elite of the time. The fan-shaped arrangement of four square impact marks—similar in size and depth and regularly spaced—could suggest a repeating firing mechanism, identified for the first time through the SCORPiò-NIDI project. Their slight angular rotation relative to a common reference axis, along with the presence of a fifth slightly misaligned mark, strengthens the hypothesis that the impacts may have been generated by a multi-shot system.

Attracting visitors, on site and/or remotely, serves as operational confirmation that the best strategy for protecting and enhancing inherited heritage is to reduce the distance between past and present. For this purpose, the strategic use of digital technologies is a powerful tool for overcoming possible cultural and institutional crises. When used wisely, technologies accelerate knowledge and enhance the effects of key factors that help meet contemporary needs. Users—elected as actors and protagonists of the simulated siege—will contribute, through their engagement online or in person, to attracting resources in support of educational, social, and tourism programs. The same archive of models will serve as a reference point for identifying a data ecosystem. A digital platform based on 3D technology (or a digital library) will be used to support the entire process of data production, storage, and management.

Digital language makes it possible to build a territory of shared assets—a framework of information in which everyone can contribute with their own expertise. The history of the siege can be rewritten, modified, or expanded based on objectively identified, classified, described, and critically used elements in line with the intended goals. Digital twins, alongside functioning physical demonstrators, if placed within reconstructed scenarios or in situ in front of the perimeter of the archeological area, can become components of museum installations for in-, on-, and off-site, or online itineraries. Digital culture refers to an interdisciplinary approach and strategy that, in accordance with the international definition of "Heritage", merges new forms of knowledge, promoting a synthesis of science, humanity, and humanism. With the "physicality" of digital constructs, multimedia and multimodal experiences are accompanied by advanced visualization to stimulate, develop, amplify, and inspire human capabilities—including socialization and collaboration that occurs through or around intelligent objects and environments.

The main problems to be addressed and resolved, in light of the objective, appear to be systemic. The relevant implications are tied to their repercussions; in the short term, the workflow must ensure accessibility and interoperability for flexibly updated digital content. In the medium and long term, it must support the promotion of updated forms of sustainable use—acting as a driving force for cultural development and local economies.

2. Objectives Achieved and in Progress

Measurement techniques, applied to the gathering of point clouds acquired with active and passive sensors, have placed a radical change in front of surveyors—a change in content rather than techniques and/or tools.

Representation based on the visualization of 3D reality has offered the opportunity to rethink the traditional cognitive–communicative process. Image- and range-based coordinates allow us to overcome the limits of the two-dimensional drawing or image, configuring a three-dimensional model that, although discontinuous, can be explored in every direction and detail in the absence of physical objects and scenarios.

Management of the acquisitions is entrusted to automatic, semi-automatic, or manual procedures to transform the coordinate points into numerically controlled models. Even today, there is still discussion about the certified reliability of the data, the optimization of the tools, the advancement of the programmed instructions for the "smart" recognition of the shape information to be inserted at the origin of the workflows.

On the other hand, studies that place the heuristic investigation of scientifically obtained results at the center of their objectives are rare. Cognitive solicitations derive from the paths or the validity of interpretative choices.

This conference aimed to discuss the limits and potentials of advanced survey and representation processes, mainly aiming to promote the debate on the solicitations derived. The responses highlighted a multiplicity of aspects that seem to reaffirm the identity of the cultural area (cultures and cultural production ERC SH5). The scientific purpose of the representation, more than a tool, appears to be a lever to discern what initially appears smooth and indistinct (*laevo*) and which instead imposes itself at the end of the emerging analysis (re).

In the order presented in this volume, the articles correspond to a work in progress. The first contribution (Russo, Rossi) is based on the exceptionality represented by the possibility of studying the indisputable ballistic impacts of the Republican era. The damage inflicted by stone balls, dart tips, and sling bullets, found today along the ancient Pompeii city walls—to be precise, between Vesuvio Gate and Ercolano Gate—appears to be "relics" as the only existing evidence in the world of the lethality of elastic torsion weapons (of which we do have a few battered archeological remains). With respect to the studies carried out, some of which are original, calculation of the ballistic parameters derived from the effects found on the extrados of the wall section is specified and extended. For this purpose, in 2002 and 2016, the authors surveyed significant cavities using direct methods. The collected data were used to calculate the volume of pulverized stone material. Parameters were drawn in the light of ancient treatises in which engineers of the time proportioned causes–effects based on experimental verification of the damage caused. The criterion of the analysis was then transcribed to the study, never addressed before, of small indentations, favoring those of a square shape, to exclude indentations produced by ball throwers or slings.

This purpose was achieved through the acquisition and non-contact processing carried out during the survey campaign in 2024, documented in the subsequent contributions 2, 3, and 4 (Bertacchi, Rossi, Formicola et al.). The purpose of these studies was to promote the debate on the results of the non-contact survey. The results raised original and innovative hypotheses regarding the existence and use of the *polybolos*, a repeating scorpion. The validation of techniques, tools, and methods is necessary, but, above all, heuristic interpretation of the results is important. Projective aspects are still situated as the basis of instructions for processing software.

Evident geometric sensitivity also appears in the solutions needed to trace the curves, explored in the fifth contribution (Gonizzi Barsanti), where a three-dimensional image is broken down, thanks to and by virtue of numerical radiology, into voxel elements needed for the mechanical characterization of models.

The sixth and seventh contributions identify numerical techniques (finite element method, FEM) to search for approximate solutions of problems described via partial differential equations, reducing the latter to a system of algebraic equations (Guagliano et al.; Minutolo et al.).

Reconstruction of the small catapult found in Xanten converges towards the final objectives but starts as a result of concrete premises and methods. The eighth contribution (Fratino and Rossi) studies and reconstructs a full-scale model to verify the coherence of the manufacturing techniques in relation to the theoretical and material sources used to evaluate their effectiveness. The process highlights the role of experimental archeology; this action fosters a dynamic process of questioning and guides the reading of signs, not only materials.

The ninth contribution (Fantini and Bertacchi) underlines, with full knowledge of the facts, an interdisciplinary approach to the theme. The article documents an integrated workflow that transforms high-resolution non-contact survey data. Combining terrestrial laser scanning and close-range photogrammetry, reliable reality-based digital assets were generated with multiple purposes. The contribution intended to combine advanced engineering methods with inclusive best practices for museums. Firstly, it provided reliable digital reconstructions of the impact cavities and corresponding stone projectiles using mesh optimization and inverse modeling tools. Secondly, it presents a series of physical prototypes—scale models and 1:1-scale printed models—designed according to the best practices for tactile exploration. Finally, it discusses the possible application of dislocated subdivision surfaces and multiscale LOD technologies to ensure both the readability of the overall shape and the preservation of key impact details.

Advanced visualization enriched with multidimensional information and incentives is shown to be a vector of significant aspects that prompt the dissemination of the outcomes. In this light, the tenth contribution (Casadei and Di Modica) deals with the possibility of organizing the development of an interactive digital platform for the visualization and exploration of these models using open-source tools. In particular, it focuses on the use of the ATON framework and Blender to demonstrate the functioning of the scorpion siege machine in the simulated environment, bringing the models to life through advanced animation tools. The objective focuses on the dissemination of a sector that is only (apparently) a niche but which for the "guardians of history and the builders of history" still has much to tell the contemporary world.

Figures 2 and 3 summarize the meaning of the interdisciplinary work carried out to date, summarizing the contribution made by each research group. In the applications, the machine models were reconstructed by Claudio Formicola, a PhD student, and inserted into a reality-based context created by Bertacchi, Gonizzi, and Rossi, edited in Lumion and Twinmotion and rendered by Veronica Casadei.

Figure 2. Comparative view of a technical drawing and its corresponding render, created in Lumion to enhance visual communication (credit: Veronica Casadei).

Figure 3. Models of the Roman scorpion and walls integrated into a virtual environment created using Twinmotion (credit: Veronica Casadei).

3. Essential Bibliography

3.1. The Sources

- IV century BC. Enea Tattico (Aeneas Tactician), *Poliorketikà*.
- IV century BC. Tucidide (Thūkydídēs), Perí toû Peloponnēsíou polémou.
- III century BC. Filone di Bisanzio (Philo of Byzantium, 280/220) *Belopoeica (βελοποιικά)* between 280/220.
- III century BC. Bitone, *Paraskeuastica—παρασκευαστικά*, between.
- I century BC. Vitruvio (Vitruvius Pollio Marcus), *De Architectura. Libri X.* 19AD. 1st ed.; Heralt, Roma 1486.
- I century AD. Giuseppe Flavio (Titus Flavius Josephus). *De bello Iudaico*, c. 75.
- I century AD. Erone di Alessandria, il Vecchio (Heron Alexandrinus), *Belopoeica*, between 212–290.
- XIII century AD. *Codex Parisinus inter supplementa Greca 607* (foll. 56r-58v), Paris: Bibl. Nationale, 1270.
- XIII century AD. Mariano di Jacopo, detto il Taccola. *De Ingeneis*, 2 books, between 1395–1453.

3.2. Exegesis of Sources Analysis of Findings and Reconstructions

- 1567. Daniele Barbaro. I dieci libri dell'architettura di M. Vitruvio tradotti & commentati da mons. Daniel Barbaro, Francesco de'Franceschi Senese et Giovanni Chrieger Alemano.
- 1572. Barocio, Francesco And Veneziano, Patrizio. Heronis Mechanici, apud Franciscum Franciscium Senensem, Venetiis, 1572.
- 1616. Baldo Bernardino abbate di Guastalla. Heronis Ctesibii Ctesibij Belopoeeca: Item Heronis vita. Francus Francis Editore.
- 1758. Galiani, Berardo. L'architettura di Marco Vitruvio Pollione tradotta e commentata dal marchese Berardo Galiani. Stamperia Simoniana, Napoli.

- 1836. Marini, Luigi. L'architettura di Vitruvio esposta in italiana favella ed illustrata con comenti e tavole cento quaranta in tre volumi. Roma: dai tipi appostamente preparati nel suo domicilio., Roma.
- 1867. Wescher, Carl. Poliorectique des Grecs. Traites Theoriques, Recits Historiques, Kessinger Publishing, Paris.
- 1877. Prou, Victor. La Chirobaliste d'Heron d'Alexandrie- In. Notices et extraits des manuscrits de la Bibliotheque Nationale et autres bibliotheques, 26, 1877.
- 1878. Landels John C. Engineering in the ancient world, University of California Press, Berkeley and Los Angeles.
- 1899. Schmidt W. Heronis Alexandrini opera. Leipzig, book I-.
- 1906. Schneider, Rudolf (1906). Herons Cheiroballistra. Mitteilungen des Kaiserlich Deutschen Archaologischen Instituts. Berlin, 168p.
- 1908. Puig Y.; Cadafalch, J. *Les Excavacions d'Empuries. Estudi De Topographia,* In *Annuari Inst. Catal.*, 1908, pp. 150–194.
- 1908. Schumacher, G. Tell El Mutesellim. Bericht Über Die 1903 Bis 1905 Mit Unterstützung SR. Majestät Des Deutschen Kaisers Und Der Deutschen Orientgesellschaft Vom Deutschen Verein Zur Erforschung Palästinas Veranstalteten Ausgrabungen; Rudolf Haupt: Leipzig, 1908.
- 1920. Diels, Hermann; Schramm, Erwin. Exzerpte aus Philons Mechanik B. VII und VII (vulgo funftes Buch). German translation, Exzerpte aus Philons Mechanik B. VII und VII (vulgo funftes Buch). Reimer, Berlin. (With German translation.) (Abhandlungen der preussischen Akademie der Wissenschaften, Philosoph.-hist. Kl. 12.
- 1929. Rehm, A.; Schramm, E. Bitons Bau von Belagerungsmaschinen und Geschützen: Griechisch und Deutsch, boos 2. Reprint: 2019. Berlin: De Gruyter Oldenbourg.
- 1969. Marsden, Eric William. Greek and Roman Artillery. Historical Development. 2nd ed.; Clarendon Oxford.
- 1971. Vitucci, Giovanni (ed. by). La guerra giudaica. Fondazione Lorenzo Valla. 3rd ed. Mondadori, Milano1995.
- 1972. Garlan, Yvon. La guerre dans l'antiquité. Paris. Trad. it. Guerra e società nel mondo antico. Bologna: Il Mulino 1985.
- 1973. Garlan, Yvon. Cite's, arme'es et strate'gie al'e'poque helle'nistique d'apre's l'oeuvre de Philonde Byzance. Historia 22 1985, pp. 16–33.
- 1974. Baatz, D. Teile spatromischen Ballistenaus Gornea und Orşova (Romania), Saalburg-Jahrbuch 31, 1974, 50–72.
- 1974. Garlan, Yvon. Le livre "V" de la "syntaxe mecanique" de Pihilon de Byzance. In: Recherches de poliorcétique grecque, Ecole Française d'Athenes De Boccard, Paris.
- 1974. Garlan, Yvon. Recherches de poliorcetique grecque. Ecole francais d'Atenes, Paris.
- 1977. Baatz, D. The Hatra Ballista, Sumer XXXIII 1977, 141–151.
- 1978. Baatz, D. Recent Finds of Ancient Artillery, Britannia 9, 1978, 1–17. Recenti Ritrovamenti Di Artiglieria Antica, Britannia.
- 1978. Gille, Bertrand. Histoire des techniques. 1st ed. eng. The History of Techniques: Gordon and Breach Science Publishers, New York, 1986, 2 books. 1st ed. It Storia delle tecniche. Editori Riuniti; 1985.
- 1981. Baatz, D, Feugere Michel. Elements d'une catapulte romaine trouvee a Lyon, Gallia 39, (1981) 201–209.
- 1984. Mariano di Jacopo, detto il Taccola. De Ingeneis 2 boos, British Library Additional 34113 Editore: L. Reichert, Wiesbaden.
- 1990. Whitehead, David (ed. by). Aineias the Tactician. How to Survive Under Siege. Clarendon Press, Oxford, 214p.

- 1994. Holley, A.E. The Ballista Balls from Masada. In Masada IV: the Yigael Yadin excavations 1963–1965. Final reports; Aviram, J., Foerster, G., Netzer, E., Eds.; Israel Exploration Society: Hebrew University of Jerusalem: Jerusalem, 1994; pp. 347–365.
- 1995. Wilkins, A.; Morgan, L. Recontrwucting cheirobbalista. Journal of Roman Military Equipment Studies, 6, 1995, pp. 5–59 and 11, 2000, pp. 77–101.
- 1997. Gros, Pierre. Vitruvio, De Architectura, Editors Corso Antonio and Romano e Elisa, Einaudi, Torino, book 2 s, 2007.
- 2000. Sulmona, Alan; Morgan L. Scorpio and Cheiroballistra, Journal of Roman Military Equipment Studies 11, 2000, 77–101.
- 2002. Russo, F.; Russo, F. Tormenta. Venti secoli di artiglieria meccanica. Roma: Stato Maggiore Esercito, Roma, 2002.
- 2003. Campbell, D.B., Greek and Roman Artillery 399 BC–AD 363, Osprey Publishing, Oxford.
- 2003. Wilkins A. Roman Artillery, Shire Publications, Princes Risborough, Bucks., United Kingdom.
- 2004. Russo, Flavio. L'artiglieria delle legioni romane. Le macchine da guerra che resero invincibile l'esercito romano. Roma: Istituto Poligrafico e Zecca dello Stato.
- 2006. Russo, Flavio. Indagine sulle Forche Caudine, Rivista Militare, Roma.
- 2006. Wilkins, A.; Barnard, H.; Rose, P.J. Roman Artillery Balls from Qasr Ibrim. The Sudan Archaeological Research Society, Egypt, Sudan and Nubia. 10, pp. 64–78.
- 2008. Russo, F.; Russo, F. Gli scorpioni della Repubblica. La catapulta di Ampurias. Edizioni Scientifiche e Artistiche, Torre del Greco, 46p.
- 2010. Trabucco, Oreste. L'opere Le opere stupende dell'arti più ingegnose, Olschki, Firenze, 20101971.
- 2012. Guzzomi A. L., Maraldi, M., Molari P.G., (2012). A historical review of the modulus concept and its relevance to mechanical engineering design today, Mechanism and MachineTheory, 50, 1–14.
- 2012. Molari, P.G.; Maraldi, M.; Angelini, G.; Bignami, S.; Lionello, G. La ricostruzione della balista di Vitruvio. In Proceedings of the Quinta Giornata di Studio Ettore Funaioli (V GEF), Bologna, 15 luglio 2011; **2012**; pp. 1–26.
- 2020. Di Marco, U.; Molari, P.G. Archimede ed il sistema di caricamento della balista da un talento utilizzata nella fortezza dell'Eurialo di Siracusa. In Proceedings of the Proceedings of the 4th International Conference | Atti dell'8° Convegno Nazionale, Naples, 2020, December 11th; D'Agostino, S., d'Ambrosio Alfano, F.R., Eds.; Cuzzolin: Napoli, **2020**; Vol. I, pp. 393–408.
- 2024. Wilkins, A. Roman Imperial Artillery: Outranging the Enemies of the Empire. 3°. Oxford: Archaeopress.

3.3. Historical-Military Insight into the Northern Route of the City of Pompeii

- 1804. Piranesi, F.; Guattani, G.A. Antiquités de la Grande Grèce, aujourd'hui Royaume de Naples, gravées par François Piranesi; Piranesi Leblanc: Paris, FR, 1804; Vol. 1.
- 1860. Fiorelli, G. Pompeianarum Antiquitatum Historia. Vol. 1. Edit. Prid. Kal. Decembris 1824. Mazois, F. Les ruines de Pompéi. Paris, FR: Imprimerie et Libraire de Firmin Didot, Paris (FR).
- 1862. Fiorelli, G. Pompeianarum Antiquitatum Historia; Edit. Prid. Non. Martias: Napoli, 1862; Vol. 2.
- 1873. Fiorelli, G. Gli scavi di Pompei dal 1861 al 1872 relazione al Ministro della istruzione pubblica di Giuseppe Fiorelli; Tipografia italiana nel liceo V. Emanuele: Napoli, 1873.
- 1917. Sogliano, A. Porte, torri e vie di Pompei nell'epoca sannitica, Atti della R. Accademia di Archeologia Lettere e Belle Arti di Napoli VI: 155–80.

- 1918. Van Buren, A.W. Studies in the Archaeology of the Forum at Pompeii. Memoirs of the American Academy in Rome 2: 67–76. https://doi.org/10.2307/4238509.
- 1925. Van Buren, A.W. Further Studies in Pompeian Archaeology, Memoirs of the American Academy in Rome 5: 103–13. https://doi.org/10.2307/4238527.
- 1929. Maiuri, A. Studi e ricerche sulla fortificazione di Pompei. In Monumenti antichi pubblicati per cura della R. Accademia Nazionale dei Lincei, 33Vol. Milano: Hoepli. 1930.
- 1943. Maiuri, Amedeo. Introduzione allo studio di Pompei, a cura di G. Oscar Onorato, Corso di Antichità Pompeiane ed Ercolanesi 1942–43. Napoli: GUF.
- 1943. Maiuri, A. L'isolamento della cinta muraria fra Porta Vesuvio e Porta Ercolano, in Notizie degli Scavi di Antichità, IV, **1943**, pp. 275–314.
- 1960. Maiuri, A. Pompei. Sterro dei cumuli e isolamento della cinta murale. Contribuito all'urbanistica della città dissepolta. Bollettino d'Arte del Ministero della Pubblica Istruzione I–II (gennaio-giugno), 166–79.
- 1985. Manning, W.H. Catalogue of the Romano-British Iron Tools, Fittings and Weapons in the British Museum. London: Published for the Trustees of the British Museum by British Museum Publications.
- 2001. Jacobelli, L. Pompei fuori le mura: note sulla gestione e l'organizzazione dello spazio pubblico e privato. In Pompei tra Sorrento e Sarno, a cura di F. Senatore. Roma: Bardi Editore. 29–61.
- 2001. Simonelli, A. Considerazioni sull'origine, la natura e l'evoluzione del Pomerium. Aevum 75(1): 119–62.
- 2004. Russo F. La cerchia di Pompei e le sue cicatrici, L'Universo rivista edita da Istituto Geografico Militare IGM a partire dal 1920. 2004.
- 2005. Russo, F.; Russo, F. 89 a.C.: assedio a Pompei: la dinamica e le tecnologie belliche della conquista sillana di Pompei. 2005 Flavius Edizioni.
- 2006. García y García, L. Danni di guerra a Pompei. Una dolorosa vicenda quasi dimenticata. Con numerose notizie sul «Museo Pompeiano» distrutto nel 1943. Studi della Soprintendenza archeologica di Pompei; L'Erma di Bretschneider: Roma, 2006; ISBN 88-8265-369-2.
- 2007. Treré, C. The walls and gates, in the world of Pompeii, a cura di J.J. Dobbins e P.W. Foss, London-New Yok, Routledge 2007, p. 142.
- 2013. de Gennaro, M.; Calcaterra, D.; Langella A. Le Pietre Storiche della Campania dall'oblio alla riscoperta, Napoli: Luciano Editore, pp. 155–77.
- 2014. Russo, V. Intorno all'antico. Conservazione e fruizione delle mura di Pompei» in Pompei accessibile. Per una fruizione ampliata del sito archeologico, Storia della tecnica edilizia e restauro dei monumenti, a cura di R. Picone. Roma: «L'Erma» di Bretschneider, **2014**, 105–17.
- 2015. Anniboletti, L. Le fasi delle fortificazioni di Pompei. Stato della conoscenza. In "Siris" 15, **2015**.
- 2015. Burns, M. Pompeii under Siege: A Missile Assemblage from the Social War. JRMES **2015**, 14/15, 1–10.
- 2015. Fabbri, Marco. 2015. Nuove ricerche per una rilettura delle mura di Pompei, in "Siris" 15, **2015**, pp. 29–47.
- 2016. Fabbri, M. Nuove ricerche per una rilettura delle mura di Pompei. Studi e Ricerche della Scuola di Specializzazione in Beni Archeologici di Matera 15, **2015**, 29–47. http://doi.org/10.4475/782.
- 2018. Osanna, M.; Picone, R. Restaurando Pompei. Riflessioni a margine del Grande Progetto. Roma: L'Erma di Bretschneider, Roma. 2010. Trabucco, O. «L'opere stupende dell'arti più ingegnose». La recezione degli Πνευματικά di Erone Alessandrino nella cultura italiana del Cinquecento, Olschki, Firenze, 2010.

- 2019. Van der Graaff, I. The Fortifications of Pompeii and Ancient Italy. NY: Routledge.
- 2020. Giglio, M.; Pesando. F. La Guerra sociale, l'assedio e la fine di Pompei sannitica. In Pompei 79 d.C. Una storia romana. Roma, Colosseo 6.11.2020–31.1.2021, a cura di M. Torelli. Milano: Electa, **2020**, 144–51.
- 2020. Vitagliano, G.; Angelone, G. Just West of Pompei. Il sito archeologico e i bombardamenti dell'estate 1943. Lecce: YCP.
- 2021. Fabbri, M. The city-wall in the Roman Age: The case study of Pompeii. Construction History 36(2), **2021**, 1–20.
- 2021. Fabbri, M.; Ducatelli, V.; Zabotti. F. Le fortificazioni di Pompei. Nuove indagini in prossimità della Torre XI detta di Mercurio. In Studi e ricerche del Parco archeologico di Pompei. Ricerche e scoperte a Pompei: in ricordo di Enzo Lippolis. Vol. 45, **2021**, a cura di M. Osanna. Roma: L'Erma di Bretschneider, 73–92.
- 2021. Vitagliano, G. Bombs on Pompeii. Academia Letters 349. https://doi.org/10.20935/AL349.

3.4. On the Digital Survey of Ballistae and Scorpions in Pompeii

- 2024a. Rossi, A. The Survey of the Ballistic Imprints for a Renewed Image of Unearthed Pompeii. *Nexus Network Journal* 26(2), **2024**, 307–24. https://doi.org/10.1007/s00004-023-00762-9.
- 2024b. Rossi, A.; Gonizzi Barsanti, S.; Bertacchi, S. Natural or anthropic? Measurement and visualisation of wall cavities in city walls. pp. 1957–78 Measure / Out of Measure. *Transitions Proceedings of the 45th International Conference of Representation Disciplines Teachers.* 2024, Franco Angeli, Milano. https://doi.org/10.3280/oa-1180-c569.
- 2024c. Bertacchi, S.; Gonizzi Barsanti, S.; and Rossi, A. Geometry of Wall Degradation: Measuring and Visualising Impact Craters in the Northern Walls of Pompeii. *SCIRES-IT-SCIentific RESearch and Information Technology* 14(1), **2024**, 111–28. https://doi.org/10.2423/i22394303v14n1p111.
- 2024d. Rossi, A.; Gonizzi Barsanti, S.; Bertacchi, S. Use of Polybolos on the City Walls of Ancient Pompeii: Assessment on the Anthropic Cavities. *Nexus Netw J.* **2024**. https://doi.org/10.1007/s00004-024-00803-x.
- 2025a. Rossi, A.; Bertacchi, S.; Formicola, C.; Gonizzi Barsanti, S. Piccole indentazioni antropiche rinvenute nella riesumata cinta urbica di Cornelia Veneria Pompeii. *Disegnare, Idee, Immagini*, 69, **2025**, 54–67. https://doi.org/10.61020/11239247-202469-06.
- 2025b: Rossi, A.; Gonizzi Barsanti, S.; Bertacchi, S. *Conoscere Pompei. Testimonianze balistiche sillane calchi digitali*, 1st ed; libreriauniversitaria.it: Limena, Italy, 2025, https://hdl.handle.net/11591/558005.

3.5. On the Criteria and Methods of Digital Survey

- 2025. Fico, D., Rizzo, D., (Eds.). *Conservation Methodologies and Practices for Built Heritage*; MDPI—Multidisciplinary Digital Publishing Institute, Basel, Switzerland.
- 2024. Bertacchi, S.; Juan Vidal, F.; Fantini, F. Ancient Architectural Design Interpretation: A Framework Based on Alexandrian Manuals. *Acta IMEKO* **2024**, 13, 1–9. https://doi.org/10.21014/actaimeko.v13i2.1833.
- 2024. Diao, H.; Jiang, X.; Fan, Y.; Li, M.; Wu, H. 3D Face Reconstruction Based on a Single Image: A Review. *IEEE Access* **2024**, 12, 59450–59473, https://doi.org/10.1109/ACCESS.2024.3381975.
- 2024. Eramo, E.; Cinque, G.E. The Vault of the So-Called Serapeum: An Ellipsoidal Geometry at Hadrian's Villa. In *Proceedings of the Graphic Horizons*; Hermida González, L., Xavier, J.P., Pernas Alonso, I., Losada Pérez, C., Eds.; Springer Nature Switzerland: Cham, **2024**; pp. 51–58. https://doi.org/10.1007/978-3-031-57579-2_7.

- 2024. Fascia, R.; Barbieri, F.; Gaspari, F.; Ioli, F.; Pinto, L. From 3D Survey to Digital Reality of a Complex Architecture: A Digital Workflow for Cultural Heritage Promotion. *The International Archives of the Photogrammetry, Remote Sensing and Spatial Information Sciences* **2024**, XLVIII-2-W4-2024, 205–212. https://doi.org/10.5194/isprs-archives-XLVIII-2-W4-2024-205-2024.
- 2024. Foschi, R.; Fallavollita, F.; Apollonio, F.I. Quantifying Uncertainty in Hypothetical 3D Reconstruction-A User-Independent Methodology for the Calculation of Average Uncertainty. *Heritage* **2024**, 7, 4440–4454, https://doi.org/10.3390/heritage7080209.
- 2024. Giordano, A.; Russo, M.; Spallone, R. a c. di. *Beyond Digital Representation: Advanced Experiences in AR and AI for Cultural Heritage and Innovative Design*. Cham: Springer Nature Switzerland.
- 2024. Rossi, A.; Cipriani, L.; Cabezos-Bernal, P.M. (eds.). Aa.Vv. 3D Digital Models. Accessibility and Inclusive Fruition. *Disegnarecon* n.17 (32), **2024**, https://doi.org/10.20365/disegnarecon.32.2024.ed.
- 2024. Sdegno, A.; Cabezos-Bernal, P.M. Oblique Analog Models, *diségno*, **2024**, no. 14, pp. 7–21, https://doi.org/10.26375/disegno.14.2024.2.
- 2024. Tesema, K.W.; Hill, L.; Jones, M.W.; Ahmad, M.I.; Tam, G.K.L. Point Cloud Completion: A Survey. *IEEE Transactions on Visualization and Computer Graphics* **2024**, 30, 6880–6899. https://doi.org/10.1109/TVCG.2023.3344935.
- 2024. Grazianova, M.; Mesaros, P. Cultural Heritage Management from Traditional Methods to Digital Systems: A Review from Bim to Digital Twin. E3S *Web of Conf.* **2024**, 550, 01015. https://doi.org/10.1051/e3sconf/202455001015.
- 2024. Calandriello, A.; D'Acunto, G.; Gigliotti, G.C. Tactile Translations: Algorithmic Modelling for Museum Inclusiveness. In *Proceedings of the Graphic Horizons*; Hermida González, L., Xavier, J.P., Amado Lorenzo, A., Fernández-Álvarez, Á.J., Eds.; Springer Nature Switzerland: Cham, **2024**; pp. 308–315.
- 2024. Nigro, L.; Montanari, D.; Sabatini, S.; De Giuseppe, M.; Benedettucci, F.M.; Lucibello, S.; Fattore, L.; Trebbi, L.; Nejat, B.; Rinaldi, T. Caress the Pharaoh. The Tactile Reproduction of Ramses II's "Mummy" in the Sapienza University Museum of the Near East, Egypt and Mediterranean. *Journal of Cultural Heritage* **2024**, 67, 158–163. https://doi.org/10.1016/j.culher.2024.02.010.
- 2024. Mousavi, Y.; Gharineiat, Z.; Karimi, A.A.; McDougall, K.; Rossi, A.; Gonizzi Barsanti, S. Digital Twin Technology in Built Environment: A Review of Applications, Capabilities and Challenges. *Smart Cities* **2024**, 7, 2594–2615. https://doi.org/10.3390/smartcities7050101.
- 2023. Clini, P.; Angeloni, R.; D'Alessio, M.; Quarchioni, R. Enhancing Onsite and Online Museum Experience Through Digital Reconstruction and Reproduction: The Raphael and Angelo Colocci Temporary Exhibition. *SCIRES-IT—SCIentific RESearch and Information Technology* 13(2), **2023**, 71–84. https://doi.org/10.2423/i22394303v13n2p71.
- 2023. Grasso, N.; Spadavecchia, C.; Di Pietra, V.; Belcore, E. LiDAR and SfM-MVS Integrated Approach to Build a Highly Detailed 3D Virtual Model of Urban Areas: In *Proceedings of the Proceedings of the 9th International Conference on Geographical Information Systems Theory, Applications and Management*; SCITEPRESS—Science and Technology Publications: Prague, Czech Republic, **2023**; pp. 128–135.
- 2023. Yang, S.; Miaole, H.; Songnian L. Three-Dimensional Point Cloud Semantic Segmentation for Cultural Heritage: A Comprehensive Review. *Remote Sensing* 15(3), **2023**, 548. https://doi.org/10.3390/rs15030548.
- 2023. Menaguale, O. Digital Twin and Cultural Heritage—The Future of Society Built on History and Art. In *The Digital Twin*; Crespi, N., Drobot, A.T., Minerva, R., Eds.; Springer International Publishing: Cham, 2023; pp. 1081–1111 ISBN 978-3-031-21343-4.

- 2023. Luther, W.; Baloian, N.; Biella, D.; Sacher, D. Digital Twins and Enabling Technologies in Museums and Cultural Heritage: An Overview. *Sensors* **2023**, 23, 1583. https://doi.org/10.3390/s23031583.
- 2023. Grieves, M.W. Digital Twins: Past, Present, and Future. In *The Digital Twin*; Crespi, N., Drobot, A.T., Minerva, R., Eds.; Springer International Publishing: Cham, **2023**; pp. 97–121 ISBN 978-3-031-21343-4.
- 2023. Kantaros, A.; Ganetsos, T.; Petrescu, F.I.T. Three-Dimensional Printing and 3D Scanning: Emerging Technologies Exhibiting High Potential in the Field of Cultural Heritage. *Applied Sciences* **2023**, 13, 4777, https://doi.org/10.3390/app13084777.
- 2023. De Luca, V.; Gatto, C.; Liaci, S.; Corchia, L.; Chiarello, S.; Faggiano, F.; Sumerano, G.; De Paolis, L.T. Virtual Reality and Spatial Augmented Reality for Social Inclusion: The "Includiamoci" Project. *Information* **2023**, 14, 38, https://doi.org/10.3390/info14010038.
- 2022. Gabellone, F. Digital Twin: A New Perspective for Cultural Heritage Management and Fruition. *Acta IMEKO* **2022**, 11, 7 pp.-7 pp., https://doi.org/10.21014/acta_imeko.v11i1.1085.
- 2022. Bevilacqua, M.G.; Russo, M.; Giordano, A.; Spallone, R. 3D Reconstruction, Digital Twinning, and Virtual Reality: Architectural Heritage Applications. In *Proceedings of the 2022 IEEE Conference on Virtual Reality and 3D User Interfaces Abstracts and Workshops* (VRW); March **2022**; pp. 92–96.
- 2021. Apollonio, F.I.; Fantini, F.; Garagnani, S.; Gaiani, M. A Photogrammetry-Based Workflow for the Accurate 3D Construction and Visualization of Museums Assets. *Remote Sensing* 13(3), **2021**, 486. https://doi.org/10.3390/rs13030486.
- 2020. Costantino, C.; Prati, D.; Predari, G.; Bartolomei, C. 3D Laser Scanning Survey for Cultural Heritage. A Flexible Methodology to Optimize Data Collection. *The International Archives of the Photogrammetry, Remote Sensing and Spatial Information Sciences* **2020**, *XLIII-B2-2020*, 821–828. https://doi.org/10.5194/isprs-archives-XLIII-B2-2020-821-2020.
- 2020. Pollard, N. *Bombing Pompeii. Ann Arbor* (USA): University of Michigan Press.
- 2020. Guidi, G.; Frischer, B.D. 3D Digitization of Cultural Heritage. In *3D Imaging, Analysis and Applications*; Liu, Y., Pears, N., Rosin, P.L., Huber, P., Eds.; Springer International Publishing: Cham, **2020**; pp. 631–697 ISBN 978-3-030-44070-1.
- 2019. Gaiani, M.; Apollonio, F.I.; Fantini, F. Evaluating Smartphones Color Fidelity and Metric Accuracy for the 3D Documentation of Small Artifacts. *The International Archives of the Photogrammetry, Remote Sensing and Spatial Information Sciences* **2019**, XLII-2-W11, 539–547. https://doi.org/10.5194/isprs-archives-XLII-2-W11-539-2019.
- 2019. Sable, U.; Borlepwar, P.T. Recent Developments in the Field of Rapid Prototyping: An Overview. In *Proceedings of the Proceedings of International Conference on Intelligent Manufacturing and Automation*; Vasudevan, H., Kottur, V.K.N., Raina, A.A., Eds.; Springer: Singapore, **2019**; pp. 511–519.
- 2018. Apollonio, F.I.; Basilissi, V.; Callieri, M.; Dellepiane, M.; Gaiani, M.; Ponchio, F.; Rizzo, F.; Rubino, A.R.; Scopigno, R.; Sobra', G. A 3D-centered information system for the documentation of a complex restoration intervention. *Journal of Cultural Heritage* 29, **2018**, 89–99. https://doi.org/10.1016/j.culher.2017.07.010.
- 2018. Apollonio, F.I.; Bertacchi, S.; Bertacchi, G.; Ballabeni, M.; Torello, M.; Montanari, R.; Saragoni, L. *Progetto Sacher. Piattaforma Cloud per i Beni Culturali e servizi integrati per il restauro. REC Recupero & Conservazione*, 2018, 68–75.
- 2018. Attenni, M.; Bartolomei, C.; Inglese, C.; Ippolito, A.; Morganti, C.; Predari, G. Low Cost Survey and Heritage Value. *SCIRES-IT—SCIentific RESearch and Information Technology* 7(2), **2018**, 115–32. https://doi.org/10.2423//i22394303v7n2p115.

- 2017. Adembri, B.; Cipriani, L.; Bertacchi, G. Guidelines for a Digital Reinterpretation of Architectural Restoration Work: Reality-Based Models and Reverse Modelling Techniques Applied to the Architectural Decoration of the Teatro Marittimo, Villa Adriana. *ISPRS—International Archives of the Photogrammetry, Remote Sensing and Spatial Information Sciences* **2017**, XLII-5/W1, 599–606, https://doi.org/10.5194/isprs-archives-XLII-5-W1-599-2017.
- 2017. Apollonio, F.I.; Rizzo, F.; Bertacchi, S.; Dall'Osso, G.; Corbelli, A.; Grana, C. SACHER: Smart Architecture for Cultural Heritage in Emilia Romagna. In *Digital Libraries and Archives*. IRCDL 2017. Vol. 733, Communications in Computer and Information Science, a cura di C. Grana e L. Baraldi. Cham: Springer, **2017**, 142–56.
- 2017. Cipriani, L.; Fantini, F. Digitalization Culture vs Archaeological Visualization: Integration of Pipelines and Open Issues. *The International Archives of the Photogrammetry, Remote Sensing and Spatial Information Sciences* **2017**, *XLII-2-W3*, 195–202, https://doi.org/10.5194/isprs-archives-XLII-2-W3-195-2017.
- 2017. Gaiani, M. Management and Communication of Archaeological Artefacts and Architectural Heritage Using Digital IS. What Today? What Next? *Archeologia e Calcolatori* XXVIII (2), **2017**. https://doi.org/10.19282/AC.28.2.2017.34.
- 2017. Rossi, A.; Nanetti, A. a c. di. *Heritage buildings conservation: methods and techniques. Atti delle giornate di studio* (Napoli, 28–29 luglio 2015). Padova: libreriauniversitaria.it.
- 2017. Balletti, C.; Ballarin, M.; Guerra, F. 3D Printing: State of the Art and Future Perspectives. *Journal of Cultural Heritage* **2017**, *26*, 172–182. https://doi.org/10.1016/j.culher.2017.02.010.
- 2016. Clini, P.; Frapiccini, N.; Mengoni, M.; Nespeca, R.; Ruggeri, R. SfM Technique and Focus Stacking for Digital Documentation of Archaeological Artifacts. *The International Archives of the Photogrammetry, Remote Sensing and Spatial Information Sciences XLI-B5* (XXIII ISPRS Congress, 12–19 July 2016, Prague, Czech Republic), **2016**, 229–36. https://doi.org/10.5194/isprs-archives-XLI-B5-229-2016.
- 2016. Gaiani, M.; Remondino, F.; Apollonio, F.I.; Ballabeni, A. An Advanced Pre-Processing Pipeline to Improve Automated Photogrammetric Reconstructions of Architectural Scenes. *Remote Sensing* 8(3), **2016**, 178. https://doi.org/10.3390/rs8030178.
- 2016. Pintus, R.; Kazim P.; Ying Yang, T.W.; Gobbetti, E.; Rushmeier. H. A Survey of Geometric Analysis in Cultural Heritage. *Computer Graphics Forum* 35(1), **2016**, 4–31. https://doi.org/10.1111/cgf.12668.
- 2014. Apollonio, F.I.; Ballabeni, A.; Gaiani, M.; Remondino, F. Evaluation of Feature-Based Methods for Automated Network Orientation. The *International Archives of the Photogrammetry, Remote Sensing and Spatial Information Sciences* XL–5, 2014, 47–54. https://doi.org/10.5194/isprsarchives-XL-5-47-2014.
- 2014. Cipriani, L.; Fantini, F.; Bertacchi, S. 3D Models Mapping Optimization through an Integrated Parameterization Approach: Cases Studies from Ravenna. *ISPRS—International Archives of the Photogrammetry, Remote Sensing and Spatial Information Sciences. ISPRS Technical Commission V Symposium*—25 June **2014**, Riva Del Garda, Italy XL–5, 2014, 173–80. https://doi.org/10.5194/isprsarchives-XL-5-173-2014.
- 2012. Fantini, F. Modelos con nivel de detalle variable realizados mediante un levantamiento digital aplicados a la arqueología. *EGA Expresión Gráfica Arquitectónica* **2012**, 306–317. https://doi.org/10.4995/ega.2012.1383.
- 2011. Guidi, G.; Russo, M. Reality-Based and Reconstructive Models: Digital Media for Cultural Heritage Valorization. *SCIRES-IT—SCIentific RESearch and Information Technology* 1(2), **2011**, 71–86. https://doi.org/10.2423/i22394303v1n2p71.

- 2010. Fassi, F.; Achille, C.; Fregonese, L.; Monti, C. Multiple Data Source for Survey and Modelling of Very Complex Architecture. *International Archives of Photogrammetry, Remote Sensing and Spatial Information Sciences* **2010**, *XXXVIII*, 234–239.
- 2005. Scolari, M. *Il disegno obliquo. Una storia dell'antiprospettiva*; Marsilio Editori: Venezia, 2005; ISBN 978-88-317-8617-1.
- 2004. Guarnieri, A.; Vettore, A.; El-Hakim, S.; Gonzo, L. Digital Photogrammetry and Laser Scanning in Cultural Heritage Survey. In *Proceedings of the ISPRS XX. Symposium*, Com. V., WG; 15 June **2004**; Vol. 2, pp. 1–6.
- 2004. Li, Q.; Griffiths, J.G. Least Squares Ellipsoid Specific Fitting. In *Proceedings of the Proceedings of the Geometric Modeling and Processing* **2004** (GMP'04); April 2004; pp. 335–340.
- 2004. Migliari, R. Per una Teoria del rilievo architettonico. In *Disegno come modello: riflessioni sul disegno nell'era informatica*, a cura di R. Migliari. **2004**, Roma: Kappa, 63–65.
- 2004. Sardo, N. *La figurazione plastica dell'architettura. Modelli e rappresentazione*; Kappa Edizioni: Roma, 2004.
- 1997. *Las Casas Del Alma: Maquetas Arquitectónicas de La Antigüedad* (5500 a.C./300 d.C.); Azara, P., Ed.; CCCB Centre de Cultura Contemporània de Barcelona: Barcelona, España,.
- 1994. Millon, H.A. *I modelli architettonici nel Rinascimento. In Rinascimento: da Brunelleschi a Michelangelo. La rappresentazione dell'architettura*; Millon, H.A., Magnago Lampugnani, V., Eds.; Bompiani: Milano, 1994; pp. 19–72 ISBN 88-452-2266-7.
- 1993. Gavinelli, C. *Storie di modelli esibitivi e critici: modelli storico-critici di rappresentazione oggettuale di visualizzazione interpretativa*; Alinea, Firenze 2002. Millon, H., Munshower, S.S., Eds.; 1st edition.; Penn State Department of Art History: University Park, Pa.
- 1988. Millon, H.A.; Smyth, C.H. *Michelangelo Architetto. La facciata di San Lorenzo e la cupola di San Pietro*; Olivetti: Milano, 1988.
- 1966. Alberti, L.B. *L'architettura (De re aedificatoria)*; Edizioni Il Polifilo: Milano, 1966; ISBN 978-88-7050-101-8.

3.6. Reverse Engineering

- 2025. Shahab S. S, Himeur, Y.; Kheddar, H.; Amira, A.; Fadli, F.; Atalla, S.; Copiaco, A.; Mansoor, W. Advancing 3D point cloud understanding through deep transfer learning: A comprehensive survey, *Information Fusion*, Volume 113, **2025**, 102601.
- 2025. Zu, X.; Gao, C.; Liu, Y.; Zhao, Z.; Hou, R.; Wang, Y. Machine intelligence for interpretation and preservation of built heritage, *Automation in Construction*, Volume 172, **2025**, 106055 .
- 2025. Wang, Y.; Bi, W.; Liu, X.; Wang, Y. Overcoming single-technology limitations in digital heritage preservation: A study of the LiPhoScan 3D reconstruction model, *Alexandria Engineering Journal*, Volume 119, **2025**, Pages 518–530 .
- 2024. Gonizzi Barsanti, S.; Marini, M.R.; Malatesta, S.G.; Rossi, A. Evaluation of Denoising and Voxelization Algorithms on 3D Point Clouds. *Remote Sens.* **2024**, *16*, 2632. https://doi.org/10.3390/rs16142632.
- 2024. Muzahid, A.A.M.; Han, H.; Zhang, Y.; Dawei L.; Zhang, Y.; Jamshid, J. Ferdous Sohel, Deep learning for 3D object recognition: A survey, *Neurocomputing*, Volume 608, **2024**, 128436.
- 2024. Liu, D.; Cao, K.; Tang, Y.; Zhang, J.; Meng, X.; Ao, T.; Zhang, H.; Study on weathering corrosion characteristics of red sandstone of ancient buildings under the perspective of non-destructive testing, *Journal of Building Engineering*, Volume 85, **2024**,108520.

- 2024. Gao, Y.; Li, H.; Fu, W.; Chai, C.; Su, T. Damage volumetric assessment and digital twin synchronization based on LiDAR point clouds, *Automation in Construction*, Volume 157, **2024**, 105168, ISSN 0926-5805 .
- 2024. Nourian, P.; Azadi, S. Voxel graph operators: Topological voxelization, graph generation, and derivation of discrete differential operators from voxel complexes, *Advances in Engineering Software*, Volume 196, **2024**, 103722.
- 2023. Mahmoud, A.; Hu, J.S.; Waslander, S.L. Dense voxel fusion for 3D object detection. In *Proceedings of the IEEE/CVF Winter Conference on Applications of Computer Vision and Pattern Recognition*, Vancouver, BC, Canada, 17–24 June **2023**; pp. 663–672.
- 2023. Shrout, O.; Ben-Shabat, Y.; Tal, A. GraVoS: Voxel Selection for 3D Point-Cloud Detection. In *Proceedings of the IEEE/CVF Winter Conference on Applications of Computer Vision and Pattern Recognition*, Vancouver, BC, Canada, 17–24 June **2023**; pp. 21684–21693.
- 2023. Zhang, Z.; Yao, W.; Li, Y.; Zhou, W.; Chen, X. Topology optimization via implicit neural representations, *Computer Methods in Applied Mechanics and Engineering*, Volume 411, **2023**, 116052.
- 2023. Zhao, Y.; Liu, Y.; Xu, Z. Statistical learning prediction of fatigue crack growth via path slicing and re-weighting, *Theoretical and Applied Mechanics Letters*, Volume 13, Issue 6, **2023**, 100477.
- 2022. Sun, J.; Ji, Y.M.;Wu, F.; Zhang, C.; Sun, Y. Semantic-aware 3D-voxel CenterNet for point cloud object detection. *Comput. Electr. Eng.* **2022**, 98, 107677.
- 2022. He, C.; Li, R.; Li, S.; Zhang, L. Voxel set transformer: A set-to-set approach to 3D object detection from point clouds. In *Proceedings of the IEEE/CVF Conference on Computer Vision and Pattern Recognition*, New Orleans, LA, USA, 18–24 June **2022** pp. 8417–8427.
- 2022. Lv, C.; Lin, W.; Zhao, B. Voxel Structure-Based Mesh Reconstruction From a 3D Point Cloud. *IEEE Trans. Multimed.* **2022**, 24, 1815–1829.
- 2022. Goto, M.; Abe, O.; Hagiwara, A.; Fujita, S.; Kamagata, K.; Hori, M.; Aoki, S.; Osada, T.; Konishi, S.; Masutani, Y.; et al. Advantages of Using Both Voxel- and Surface-based Morphometry in Cortical Morphology Analysis: A Review of Various Applications, Magnetic Resonance. *Med. Sci.* **2022**, 21, 41–57.
- 2022. Do˘gan, S.; Güllü, H. Multiple methods for voxel modeling and finite element analysis for man-made caves in soft rock of Gaziantep. *Bull. Eng. Geol. Environ.* **2022**, 81, 23.
- 2022. Di Angelo, L.; Di Stefano, P.; Guardiani, E.; A review of computer-based methods for classification and reconstruction of 3D high-density scanned archaeological pottery, *Journal of Cultural Heritage*, Volume 56, **2022**, Pages 10–24.
- 2022. Cakir, F.; Kucuk, S. A case study on the restoration of a three-story historical structure based on field tests, laboratory tests and finite element analyses, *Structures*, Volume 44, **2022**, 1356–1391.
- 2021. Irfan, M.A.; Magli, E. Exploiting color for graph-based 3d point cloud denoising, *Journal of Visual Communication and Image Representation* **2021** 75 103027, Elsevier, Amsterdam, The Netherlands.
- 2021. Xu, Z.; Foi, A. Anisotropic denoising of 3D point clouds by aggregation of multiple surface-adaptive estimates, *IEEE Transactions on Visualization and Computer Graphics* **2021** 27 (6) 2851–2868.
- 2021. Rakotosaona, M.J.; La Barbera, V.; Guerrero, P.; Mitra, N.J.; Ovsjanikov, M. Pointcleannet: Learning to denoise and remove outliers from dense point clouds. In *Computer graphics forum*, **2021** 39, No. 1, pp. 185–203, Wiley, New York, US.

- 2021. Deng, J.; Shi, S.; Li, P.; Zhou, W.; Zhang, Y.; Li, H. Voxel r-cnn: Towards high performance voxel-based 3D object detection. *Proc. AAAI Conf. Artif. Intell.* **2021**, 35, 1201–1209.
- 2021. Xu, Y.; Tong, X.; Stilla, U. Voxel-based representation of 3D point clouds: Methods, applications, and its potential use in the construction industry, *Automation in Construction*, Volume 126, **2021**, 103675.
- 2021. Shitong, L.; Hu, W. Score-based point cloud denoising Proceedings of the *IEEE/CVF International Conference on Computer Vision*, Montreal, QC, Canada, **2021** 4563–4572.
- 2020. Dinesh, C.; Cheung, G.; Baji'c, I.V. Point cloud denoising via feature graph Laplacian regularization, *IEEE Transactions on Image Processing* **2020** 29 4143–4158.
- 2020. He, C.; Zeng, H.; Huang, J.; Hua, X.S.; Zhang, L. Structure Aware Single-Stage 3D Object Detection from Point Cloud. In *Proceedings of the IEEE/CVF Conference on Computer Vision and Pattern Recognition* (CVPR), Seattle, WA, USA, 13–19 June **2020**; pp. 11870–11879.
- 2020. Sas, A.; Ohs, N.; Tanck, E.; van Lenthe, G.H. Nonlinear voxel-based finite element model for strength assessment of healthy and metastatic proximal femurs. *Bone Rep.* **2020**, 12, 100263.
- 2019. Babich, M.; Kublanov, V. Voxel Based Finite Element Method Modelling Framework for Electrical Stimulation Applications Using Open-Source Software. In *Proceedings of the Ural Symposium on Biomedical Engineering, Radioelectronics and Information Technology (USBEREIT)*, Yekaterinburg, Russia, 25–26 April **2019**; pp. 127–130.
- 2019. Xue, F.; Lu, W.; Webster, C.J.; Chen, K. A derivative-free optimization-based approach for detecting architectural symmetries from 3D point clouds, *ISPRS Journal of Photogrammetry and Remote Sensing*, Volume 148, **2019**, Pages 32–40.
- 2018. Chen, H.; Shen, J. Denoising of point cloud data for computer-aided design, engineering, and manufacturing, Engineering with Computers **2018** 34 523–541, Springer Nature, London, UK.
- 2018. Han, X.; Jin, J.S.; Wang, M.; Jiang, W. Guided 3D point cloud filtering, *Multimedia tools and Applications* **2018,** 77 17397–17411, Springer nature, London, UK.
- 2018. Zhou, Q. Y.; Park, J.; Koltun, V. Open3D: A modern library for 3D data processing. *arXiv preprint arXiv* **2018**, 1801.09847.
- 2017. Digne, J.; Franchis, C.D. *The bilateral filter for point clouds*, Image Processing online 278–287, Centre Borelli, ENS Paris-Saclay, France 2017.
- 2017. Sapozhnikov, S.B.; Shchurova, E.I. Voxel and Finite Element Analysis Models for Ballistic Impact on Ceramic-polymer Composite Panels. *Procedia Eng.* **2017**, 206, 182–187.
- 2017. Baert J., Cuda voxelizer: A gpu-accelerated mesh voxelizer. https://github.com/Forceflow/cuda_voxelizer, 2017.
- 2013. He, K.; Sun, J.; Tang, X. Guided image filtering, *IEEE Transactions on Pattern Analysis and Machine Intelligence* **2013** 35 (6) 1397–1409.
- 2012. Watanabe, K.; Iijima, Y.; Kawano, K.; Igarashi, H. Voxel Based Finite Element Method Using Homogenization. *IEEE Trans. Magn.* **2012**, 48, 543–546.
- 1999. Lee, T.Y.; Weng, T.L.; Lin, C.H.; Sun, Y.N. Interactive voxel surface rendering in medical applications. *Comput. Med. Imaging Graph.* **1999**, 23, 193–200.
- 1998. Tomasi, C.; Manduchi, R. Bilateral filtering for gray and color images, in: *Sixth International Conference on Computer Vision*, **1998** IEEE pp. 839–846.

3.7. Finite Element Analysis

- 2024. Zona, R.; Minutolo, V. A dislocation-based finite element method for plastic collapse assessment in solid mechanics. *Archive of Applied Mechanics, 94,* **2024**, 1531–1552. https://doi.org/10.1007/s00419-024-02594-6.
- 2024. Pei, G.; Xiao, D.; Zhang, M.; Jiang, J.; Xie, J.; Li, X.; Guo, J. Study on the Dynamic Fracture Properties of Defective Basalt Fiber Concrete Materials Under a Freeze–Thaw Environment. *Materials,* 17(24), **2024**, 6275. https://doi.org/10.3390/ma17246275.
- 2023. Esposito, L.; Palladino, S.; Minutolo, V. An effective free-meshing and linear step-wise procedure to predict crack initiation and propagation. *Theoretical and Applied Fracture Mechanics, 130,* **2023**, 104240. https://doi.org/10.1016/j.tafmec.2023.104240.
- 2022. Palladino, S.; Minutolo, V.; Esposito, L. Hybrid semi-analytical calculation of the stress intensity factor for heterogeneous and functionally graded plates. *Engineering Fracture Mechanics, 274,* **2022**, 108763. https://doi.org/10.1016/j.engfracmech.2022.108763.
- 2021. Liu, K.; Zhao, J. Progressive damage behaviours of triaxially confined rocks under multiple dynamic loads. *Rock Mechanics and Rock Engineering, 54,* **2021**, 573–590. https://doi.org/10.1007/s00603-021-02408-z.
- 2021. Heap, M.J.; Violay, M.E. The mechanical behaviour and failure modes of volcanic rocks: a review. *Bull Volcanology* 83, **2021**, 33. https://doi.org/10.1007/s00445-021-01447-2.
- 2021. Zona, R.; Ferla, P.; Minutolo, V. Limit analysis of conical and parabolic domes based on semi-analytical solution. *Journal of Building Engineering.* 44, **2021**, 103271. https://doi.org/10.1016/j.jobe.2021.103271.
- 2021. Palladino, S.; Esposito, L.; Ferla, P.; Zona, R.; Minutolo, V. Functionally Graded Plate Fracture Analysis Using the Field Boundary Element Method. *Applied Sciences, 11,* **2021**, 88465. https://doi.org/10.3390/app11188465.
- 2017. Cattania, C.; Rivalta, E.; Hainzl, S.; Passarelli, L.; Aoki, Y. A nonplanar slow rupture episode during the 2000 Miyakejima dike intrusion. *Journal of Geophysical Research: Solid Earth, 122,* **2017**, 2054–2068. https://doi.org/10.1002/2016JB013722.
- 2004. James, M. R., et al. Fracture toughness measurements on igneous rocks using a high-pressure, high-temperature rock fracture mechanics cell. *Journal of Volcanology and Geothermal Research, 132*(2–3), **2004**, 159–172. https://doi.org/10.1016/S0377-0273(03)00343-3.
- 2001. Minutolo, V.; Gesualdo, A.; Nunziante, L. Local Collapse in Soft Rock Bank Cavities. *Journal of Geotechnical and Geoenvironmental Engineering.* 127 (12), **2001**, 1037–1042. https://doi.org/10.1061/(ASCE)1090-0241(2001)127:12(1037).
- 1989. Zienkiewicz, O.C., Taylor, R.L., *The Finite Element Method,* McGraw Hill, London, 1989.

3.8. Dissemination

- 2025. Apollonio, F.I.; Zannoni, M.; Fantini, F.; Garagnani, S.; Barbieri, L. Accurate Visualization and Interaction of 3D Models Belonging to Museums' Collection: From the Acquisition to the Digital Kiosk. *J. Comput. Cult. Herit.* 2025, 18, 5:1–5:25. https://doi.org/10.1145/3704812.
- 2024. Balzani, R.; Barzaghi, S.; Bitelli, G.; Bonifazi, F.; Bordignon, A.; Cipriani, L.; Colitti, S.; Collina, F.; Daquino, M.; Fabbri, F.; et al. Saving Temporary Exhibitions in Virtual Environments: The Digital Renaissance of Ulisse Aldrovandi—Acquisition and Digitisation of Cultural Heritage Objects. *Digital Applications in Archaeology and Cultural Heritage* 2024, 32, e00309. https://doi.org/10.1016/j.daach.2023.e00309.
- 2024. Brubaker, S. Realizing 3D Animation in Blender: Master the Fundamentals of 3D Animation in Blender, from Keyframing to Character Movement. Packt Publishing Ltd., 2024; pp. 4–17.
- 2024. Meier, C.; Saorín, J.L.; Díaz Parrilla, S.; Bonnet De León, A.; Melián Díaz, D. User Experience of Virtual Heritage Tours with 360° Photos: A Study of the Chapel of

Dolores in Icod de Los Vinos. *Heritage*, 2024, *7*, 2477–2490. https://doi.org/10.3390/heritage7050118.
- 2024. Casillo, M.; Colace, F.; Gaeta, R.; Lorusso, A.; Santaniello, D.; Valentino, C. Revolutionizing Cultural Heritage Preservation: An Innovative IoT-Based Framework for Protecting Historical Buildings. *Evol. Intel.* 2024, 17, 3815–3831. https://doi.org/10.1007/s12065-024-00959-y.
- 2024. Li, F.; Spettu, F.; Achille, C.; Vassena, G.; Fassi, F. The Role of Web Platforms in Balancing Sustainable Conservation and Development in Large Archaeological Site: The Naxos Case Study. *The International Archives of the Photogrammetry, Remote Sensing and Spatial Information Sciences* 2024, *XLVIII-2-W8-2024*, 2024, 303–310. https://doi.org/10.5194/isprs-archives-XLVIII-2-W8-2024-303-2024.
- 2024. Spiess, F.; Waltenspül, R.; Schuldt, H. *The Sketchfab 3D Creative Commons Collection* (S3D3C) 2024.
- 2023. Pietroni, E.; Menconero, S.; Botti, C.; Ghedini, F. e-Archeo: A Pilot National Project to Valorize Italian Archaeological Parks through Digital and Virtual Reality Technologies. *Appl. Syst. Innov.* 2023, 6, 38. https://doi.org/10.3390/asi6020038.
- 2023. Belec, A. *Photorealistic Materials and Textures in Blender Cycles: Create impressive production-ready projects using one of the most powerful rendering engines*, 4th ed.; Packt Publishing Limited, 2023; pp. 351–363.
- 2023. Fanini, B.; Pagano, A.; Pietroni, E.; Ferdani, D.; Demetrescu, E.; Palombini, A. Augmented Reality for Cultural Heritage. In *Springer Handbook of Augmented Reality*; Nee, A.Y.C., Ong, S.K., Eds.; Springer International Publishing: Cham, 2023, pp. 391–411, https://doi.org/10.1007/978-3-030-67822-7_16.
- 2023. Ferdani, D.; Ronchi, D.; Fanini, B.; Manganelli Del Fà, R.; d'Annibale, E.; Bordignon, A.; Pescarin, S. Brancacci Chapel in Florence: Surveying and Real-Time 3D Simulation for Conservation and Communication Purposes. *The International Archives of the Photogrammetry, Remote Sensing and Spatial Information Sciences* 2023, XLVIII-M-2–2023, 535–540, https://doi.org/10.5194/isprs-archives-XLVIII-M-2-2023-535-2023.
- 2023. Brusaporci, S. Architectural Heritage Imaging: When Graphical Science Meets Model Theory. *DISEGNARECON* 2023, *16*, 1–4, https://doi.org/10.20365/disegnarecon.31.2023.ed.
- 2023. Bruciati, A.; D'Alessandro, L.; Empler, T.; Fusinetti, A. VILLÆ (Tivoli, MiC). Percorsi di inclusione museale e accessibilità. In *Proceedings of the DAI—Il Disegno per l'Accessibilità e l'Inclusione. Atti del II convegno DAI*, Udine 1-2 dicembre 2023; Sdegno, A., Riavis, V., Eds.; PUBLICA: Alghero, 2023; pp. 508–521.
- 2023. Bruciati, A.; D'Alessandro, L. Passepartout. *Il museo di tutti per tutti. Un progetto di accessibilità delle VILLÆ. In Aree archeologiche e accessibilità. Riflessioni ed esperienze*; Anguissola, A., Tarantino, C., Eds.; Pisa University Press: Pisa, 2023; pp. 127–137 ISBN 978-88-3339-725-2.
- 2022. Costa, S.; Cordera, P.; Poulot, D. Storytelling? Una narrazione a più voci. In *Storytelling: esperienze e comunicazione del Cultural Heritage*; Costa, S.; Cordera, P.; Poulot, D., Eds; Bononia University Press, Bologna, Italy, 2022; pp 1–7.
- 2021. Bonacini, E.; Marangon, G. *Lo storytelling digitale partecipato come strumento didattico di divulgazione culturale.* Cuadernos de Filología Italiana, 28, 2021; pp. 405–425 https://doi.org/10.5209/cfit.70449.
- 2021. Fanini, B.; Ferdani, D.; Demetrescu, E.; Berto, S.; D'Annibale, E. ATON: An Open-Source Framework for Creating Immersive, Collaborative and Liquid Web-Apps for Cultural Heritage. *Appl. Sci.* 2021, 11, 11062 https://doi.org/10.3390/app112211062.

- 2021. Fanini, B.; Ferdani, D.; Demetrescu, E. Temporal Lensing: An Interactive and Scalable Technique for Web3D/WebXR Applications in Cultural Heritage. *Heritage* 2021, 4, 710–724, https://doi.org/10.3390/heritage4020040.
- 2021. Peinado-Santana, S.; Hernández-Lamas, P.; Bernabéu-Larena, J.; Cabau-Anchuelo, B.; Martín-Caro, J.A. Public Works Heritage 3D Model Digitisation, Optimisation and Dissemination with Free and Open-Source Software and Platforms and Low-Cost Tools. *Sustainability* **2021**, 13, 13020. https://doi.org/10.3390/su132313020.
- 2020. Astorga González, E.M.; Municio, E.; Noriega Alemán, M.; Marquez-Barja, J.M. Cultural Heritage and Internet of Things. In *Proceedings of the Proceedings of the 6th EAI International Conference on Smart Objects and Technologies for Social Good*; Association for Computing Machinery: New York, NY, USA, 14 September 2020; pp. 248–251.
- 2020. Ponchio, F.; Callieri, M.; Dellepiane, V. and R. Scopigno. Effective Annotations Over 3D Models. Computer Graphics Forum 39, no. 1, 2020; pp 89–105. https://doi.org/10.1111/cgf.13664.
- 2020. Negri, M.; Marini, G. *Le 100 parole dei musei*; Marsilio Editori: Venezia, 2020; ISBN 978-88-297-0078-3.
- 2019. Hosen, M.S.; Ahmmed, S.; Dekkati, S. Mastering 3D Modeling in Blender: From Novice to Pro. *ABC res. alert* 2019, *7*, 169–180, 10.18034/ra.v7i3.654.
- 2018. Vaglio, M. G. Lo storytelling per i beni culturali: il racconto. In *Racconti da museo: storytelling d'autore per il museo 4.0*; Dal Maso, C. Ed; Edipuglia, Bari, Italy, 2018; pp 27–51.
- 2018. Dal Maso, C. Storytelling: perché. In *Racconti da museo: storytelling d'autore per il museo 4.0*; Dal Maso, C Ed.; Edipuglia, Bari, Italy, 2018; pp 11–25.
- 2017. Viola, F. and Idone Cassone, V. *L'arte del coinvolgimento: emozioni e stimoli per cambiare il mondo*. Microscopi, Hoepli Editor: Milano, Italy, 2017; pp. 1–7.
- 2015. Gonizzi Barsanti, S.; Caruso, G.; Micoli, L.L.; Covarrubias Rodriguez, M.; Guidi, G. 3D Visualization of Cultural Heritage Artefacts with Virtual Reality Devices. *Int. Arch. Photogramm. Remote Sens. Spatial Inf. Sci.* 2015, XL-5/W7, 165–172, 10.5194/isprsarchives-XL-5-W7-165-2015.
- 2014. Zambruno S.; Vazzana A.; Orlandi M. Tecnologia, Beni Culturali e Turismo: i Tour Virtuali (Virtual Tours) come strumento per una corretta comunicazione dei Beni Culturali, *Storia e futuro* 34, 2014; pp. 1–4.
- 2014. Basso Peressut, L.; Caliari, P. *Architettura per l'Archeologia. Museografia e Allestimento*; Martinelli, C., Ed.; Prospettive Edizioni: Roma, 2014; ISBN 978-88-98563-06-7.
- 2010. *Manual de museología interactiva; Santacana i Mestre*, J., Martín Piñol, C., Eds.; Ediciones Trea: Gijón, España, 2010; ISBN 978-84-9704-531-5.
- 2009. Lugli, A. *Museologia*; Jaka Book: Milano, 2009; ISBN 978-88-16-43033-4.
- 2007. Battini, C.; Fantini, F. *Modelli tattili: una questione di rappresentazione. In Nuove immagini di monumenti fiorentini, rilievi con tecnologia laser scanner 3D*; Bini, M., Battini, C., Eds.; Alinea: Firenze, 2007; pp. 36–39 ISBN 978-88-6055-232-7.

3.9. Websites

- Council of the European Union. Council conclusions of 21 May 2014 on cultural heritage as a strategic resource for a sustainable Europe (2014/C 183/08) 2014; Available online: https://eur-lex.europa.eu/legal-content/EN/TXT/PDF/?uri=CELEX:52014XG0614(08) (accessed on on 21 March 2025).
- Drafted by the Central Institute for the Digitalization of Cultural Heritage—Digital Library of the Italian Ministry of Culture within the National Plan for the Digitalization of Cultural Heritage 2022–2026. https://digitallibrary.cultura.gov.it/linee-guida/ (accessed on on 22 March 2025).
- ATON. Available online: https://osiris.itabc.cnr.it/aton/ (Accessed on 26 March 2025).

- Blender. Available online: https://www.blender.org/ (Accessed on 26 March 2025).
- Three.js. Available online: https://threejs.org/ (Accessed on 31 March 2025).
- Khronos. Available online: https://www.khronos.org/gltf/ (Accessed on 26 March 2025).
- Blender 4.3 manual. Available online: https://docs.blender.org/manual/en/4.3/index.html (Accessed on 02 April 2025).
- Agisoft Metashape Professional. Available online: https://www.agisoft.com/ (accessed on February 2025).
- 3d Systems Geomagic Design X. Available online: https://www.3dsystems.com/software/geomagic-design-x (accessed on February 2025).
- Python. Available online: https://www.python.org (accessed on March 2025).
- Rhinoceros (Rhino 3D). Available online: https://www.rhino3d.com (accessed on February 2025).
- Grasshopper. Algorithmic Modeling for Rhino. Available online: https://www.grasshopper3d.com (accessed on March 2025).
- ChatGPT (OpenAI). Available online: https://openai.com/chatgpt (accessed on April 2025).
- Leica Geosystems Cyclone [Software]. Available online: https://leica-geosystems.com/it-it/products/laser-scanners/software (accessed on March 2025).

Funding: This work was supported by the Italian Ministry under the Project of Relevant National Interest (PRIN call D.D. no. 104/2022—code PRIN_20222RJE322022—CUP B53D23022100006, D.D. no. 1012/2023) "SCORPiò-NIDI. Comparative Analysis and Certified Reconstructions for a correct experimental archeology: Roman Scorpions and Ballistae for the Imperial mechanical culture, origin of European identity. Governance policy for the development and sustainable fruition of Cultural Heritag".

Acknowledgments: Thanks go to those who have worked in various capacities at the Ministry of Cultural Heritage since the beginning of the millennium. Starting with Mario Pagano, Superintendent for Archaeological Heritage of Molise since 2004, who, previously urged by Flavio Russo, Historical Consultant of the Army General Staff, took action to obtain authorization to proceed with the creation of silicone casts of the ballistic impacts found in the walls of Pompeii. The casts are now on display in the archaeological area of Saepinum (CB, Italy). We appreciate the interest shown in the research topic by MIBACT-SABAP-CS (0013535 of 23/10/17 CI. 19.10.38/1-9), equally supported by the Municipality of Pompeii (0051038/U of 06/11/2017), the Ministry of Defense, and other academic institutions. The authors reaffirm their gratitude to the Directorate and officials of the Archaeological Park of Pompeii, who have repeatedly authorized both access to the sites and the survey of the northern section of the city wall circuit (MIBACT-SSBA-PES PROTO-ARCH 0006225 14/04/2016 CI. 28.13.07/6; and prot. no. 9707 of 29/09/23). The most recent authorisation, updated to 28 February 2025 (PDF_1739786899487a35a63ff-f5f3-4360-9c25-7f65ef107e27), was subject to delivery of the laser scanner survey in the formats requested by the Park Research Laboratories. The delivery was made on the morning of 28 February. A few weeks later, a draft agreement for sharing the results was delivered to the legal team for review.

Conflicts of Interest: The authors declare no conflicts of interest.

engineering proceedings

MDPI

Editorial

Statement of Peer Review

Adriana Rossi

Department of Engineering, Università degli Studi della Campania Luigi Vanvitelli, 81031 Aversa, Italy; adriana.rossi@unicampania.it

In submitting conference proceedings to *Engineering Proceedings*, the Volume Editors certify to the publisher that all papers published in this volume have undergone peer review administered by the Volume Editors. Reviews were conducted by expert referees according to the professional and scientific standards expected of a proceedings journal.

- Type of peer review: Single-blind peer review
- Conference submission management system: University repository and e-mail
- Number of submissions sent for review: 12
- Number of submissions accepted: 10
- Acceptance rate (number of submissions accepted/number of submissions received): 83%
- Average number of reviews per paper: 2 (First Review) + 1 (Second Review)
- Total number of reviewers involved: 18
- Additional information on the review process: In the first review round, two reviewers are assigned to each paper to conduct a single-blind review. If both reviewers recommend rejection, the paper is rejected. Reviewers' comments are compiled and communicated to the authors for revision. In the second review round, one reviewer is assigned to assess whether the authors have adequately addressed all comments. If the revisions are found satisfactory, the paper is forwarded to the Academic Editor, who makes the final decision on its acceptance for publication in the conference proceedings, and to the Volume Editor for final communication.

Conflicts of Interest: The author declares no conflicts of interest.

Eng. Proc. **2025**, *96*, 12 https://doi.org/10.3390/engproc2025096012

engineering proceedings

MDPI

Proceeding Paper

Ancient Science: From Effects to Ballistics Parameters †

Flavio Russo [1] and Adriana Rossi [2,*]

1 Archeotecnica.com, 89122 Reggio Calabria, Italy
2 Department of Engineering, Università degli Studi della Campania Luigi Vanvitelli, Via Roma 29, 81031 Aversa, Italy
* Correspondence: adriana.rossi@unicampania.it

† Presented at the Conference "Discovering Pompeii: From Effects to Causes—From Surveying to the Reconstructions of Ballistae and Scorpiones", Aversa, Italy, 27 February 2025.

Abstract: A well-equipped legionary army prepared to lay siege to Pompeii. Among the weapons deployed along the northern stretch of the city walls were battering rams and mobile siege towers equipped with ballistae and scorpions. The impact marks from Republican-era stone balls and dart tips remain visible today between the Vesuvio and Ercolano Gates. In 2002 and 2016, the authors surveyed significant cavities using both direct and indirect methods. The collected data were then used to calculate the volume of fractured stone material. Given the hardness of the wall ashlars, ballistic parameters were quantified based on Hellenistic treatises. The results make it possible to derive dimensions for reconstructing artillery calibrated to the observed effects.

Keywords: Pompeii; ballistic impact craters; terminal ballistics; ancient weapons; Philo of Byzantium; Heron of Alexandria; elastic torsion engines; scorpions; ballistae; Sullan siege

1. Introduction

In the spring of 89 BC, Sulla's army laid siege to Pompeii, initiating its attack from the northern section of the city walls, an impressive defensive wall punctuated by three towers between the Vesuvio and the Ercolano Gates. From the edge of the moat to the slopes of the volcano, a wide plain extends, an ideal area for offensive and defensive operations [1].

This segment of the perimeter fortification is undoubtedly the most suitable for the deployment of siege machines. Moreover, it is due to this evident vulnerability that its defensive potential was increased over time, until the subsequent insertion of towers [2], armed with many ballistic artillery pieces, distributed on three levels, as the embrasures testify. Sulla's troops carefully prepared for the operations, bringing in several siege towers, or *helepolis*, many ballistae and catapults (known as scorpions) with their ammunition, placing them in protected emplacements. Having completed the preparations, at dawn on a clear summer's day, the order was given to open fire simultaneously. The signal was also understood by the defenders on the walls: "[...] preceded by menacing screeching sounds, a hail of stone balls crashed against the battlements, crushing them together with the wooden shields" [3] (III, 7.243). A hail of darts, through the many breaches, pierced all those paralyzed by the uproar and violence of the scene they were still observing.

The impacts, still visible today between the gates, evoke, in a vivid and tangible manner, the violence of the scenes reconstructed based on historical accounts [4] (II, 75), not only ancient sources. According to Amedeo Maiuri [5] (pp. 145–160), the wide walkway (approximately 12 m) was protected by a double battlement (Figure 1), higher on the inner side; a defense that was not exceptional in Hellenic military architecture as confirmed by Philo of Byzantium.

(**a**)

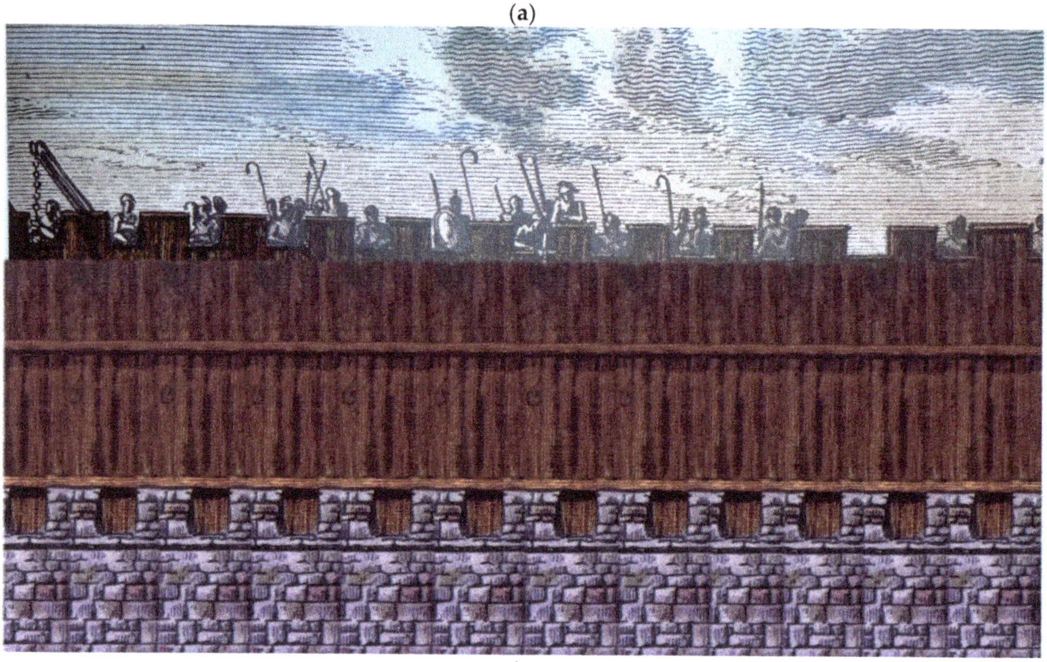

(**b**)

Figure 1. (**a**) Pompeii, original reconstruction of the walkway on both sides of Tower of Mercurio, credits Flavio Russo: (**a**) The black circles identify the larger imprints; (**b**) Protective devices placed on the battlements. Sketch based on Thucydides' description [4].

Traces of damage recognized as "imprints" caused by the impacts of projectiles launched by Sulla's artillery [5] (pp. 280–286), [6] (pp. 110–111) are still visible today along the stretch between Porta Vesuvio and Porta Ercolano (Figure 2).

Considering what we can currently observe, we explore the opportunity to derive certified data from the cavities produced by stone throwers and dart throwers on stone ashlars. Based on the scientific knowledge of the time, certified parameters of external ballistics will be functional to the proportioning of artillery, reconstructed based on the effects detected.

Figure 2. Cylindrical panorama of Pompeii's walls, northern portion. In the foreground, the remains of Tower XII. Credits: A. Rossi and M.P. Cabezos, 2016 (MIBACT-SSBA-PES PROTO-ARCH 0006225 14/04/2016 CI. 28.13.07/6)*.

2. Materials and Methods

The analysis is based on the possibility of describing in detail the geometry of the cavities caused by stone throwers or dart launchers [5,6]. Knowing the density of the stone blocks that constitute the extrados of the perimeter, the data collected will allow us to calculate the volumes of the crushed material. Basic notions of physics will allow us to quantify the kinetic energy necessary to cause the observed damage; in other words, the deformation work that allowed the penetration of stones and darts, pulverizing the material of the wall blocks. From the mechanical resistance of the impacted material, it will be possible to inversely trace the velocity of the projectiles and thus the terminal ballistics parameters. Tracing the effects back to the causes is the premise for the certified reconstruction of ballistae and scorpions used during the Sullan siege.

The connection between projectile properties and the proportioning module of siege engines was well known to the ancients, who, based on the dimensions of the hole in which the elastic bundle was tightened, calibrated ballistae and scorpions, predicting their effects: impact speed, trajectories and distances from the target were obtained based on repeated experimental tests [7] (p. 187). In the progress reports published in the Memoirs of the American Academy in Rome, Van Buren describes the anthropic traces, distinguishing them into large, medium and small [8]. It is not clear today whether the small imprints were more evident in the past; they were certainly observable before the restorations and the degradation to which the walls were subjected over the last century. The spheroidal imprints, however, remain unmistakable in shape and size, which the scholar has related to the large and medium-sized smooth stones found on site [8].

To start the investigation from the effects, it was deemed logical to begin with the survey of the large-diameter cavities that were safer to identify and measure. At the beginning of the millennium, the Superintendency of the Archaeological Park of Pompeii was asked for the required authorization to take silicone molds of the cavities of larger diameter, the most suitable to provide reliable volumes. The molds were removed in early February 2002; two layers of silicone resin were applied by the technicians one day apart and then filled with *scagliola* (Figure 3).

We now know how invasive the technique was—pigments from the plaster were removed with the molds, according to some of those present during the survey. Currently, it appears that reddish pigments can be discerned inside small and large indentations, but only a technical analysis by experts in the field could reliably confirm this (Figure 4).

(**a**) (**b**)

Figure 3. (**a**) Pompeii, silicone mold 2002. (**b**) Above left: a trace from a ballistic impact; above right, technicians at work. Below, the negative molds and positive casts currently preserved at the Saepinum area, Altilia, CB. Images courtesy of F. Russo private archive.

(**a**) (**b**)

Figure 4. Traces of colored pigments inside the anthropogenic cavities. The damage was presumably caused (**a**) by small, (**b**) large projectiles. Photo from 2016 (MIBACT-SSBA-PES PROTO-ARCH 0006225 14/04/2016 CI. 28.13.07/6)*.

Non-contact acquisition techniques promise to solve some of the problems encountered. Active or passive sensors provide reality-based acquisitions; subsequent processing creates models with inverted normals, reliable copies suitable for documenting the state of the art for future reference, offering irrefutable shared and interoperable documents in case of new natural disasters or human damage. The models can be used for research workflows, with localized analyses aimed at mechanical characterization and action management, used as a tool for scientific investigation and verification. When used as an access interface to a flexible and updatable information complex, the same 3D models can be adapted to different user levels, offering a complete integration of thematic knowledge. Reading the tangible and invisible signs, adopting a "cognitive approach" common to tactile and optical experiences, is in any case useful to retrace historical and archaeological data in order to approach the steps necessary to "build well-designed machines" as the Hero of Alexandria pointed out [7] (p. 193).

Without forgetting the involvement of the senses and emotions linked to the experience of the survey, the calibration formulas in the ancients' drawings are based on the proportional laws used by Philo and Heron. Standardizing data was an indispensable necessity on the battlefield to allow the rapid replacement of components in the event of breakage or damage, but above all to predict the lethality of the shots based on the motion of the projectile. The trajectory of a ballista's projectile, as with a catapult, is a parabola, whose flight path is divided approximately into three segments: the first is a straight shot, equal to about one-third of the entire parabola, characterized by an almost rectilinear trajectory. The second segment follows a large radius curve that directs it downwards. The third segment runs from that variation until the ground impact, which occurs after a tighter curve of the trajectory [9].

Direct, close observation of the large diameter craters identified along the Pompeian walls revealed a deep penetration into the stone ashlars of approximately 120 mm. The trajectory appears orthogonal according to the shape of the nearly "cylindrical-semispherical" trace. The direct shot, perpendicular to the extrados of the defensive wall, from a ballistic point of view, retains maximum residual kinetic energy. Assuming the distance of the target to be one hundred meters, one can conclude that the ballista artillery was placed on the ground in front of the moat and probably at the same level of the wall, or just below it, in order to hit the protective shields erected on it.

3. Result: Calibration According to Philo and Heron

The stones, being thrown along a sort of launch channel, could have been river pebbles. However, instead, the projectiles used by Sulla's ballistae in Pompeii appear so precisely worked and of a diameter classifiable in categories as described in detail by the archaeologist Van Buren. Different hypotheses have been put forward. In any case, none of them justify the standardization of diameters if the tactical reasons mentioned above are excluded—replacing broken or damaged components and foreseeing the lethality of machines with the same caliber.

All the treatise writers directly refer to the proportional relationship linking the diameter of the hole through which the "tighteners" of the twisted bundles pass with the aid of the arms used for torsion. However, no one has quantified the measurements, so reconstructions have always been based on data from the few finds discovered and identified as *modioli* of elastic torsion machines. Instead, we would like to adopt the proportional criterion recalled by Vitruvius (*De Arch.*, X, 10–13, 19 AD) for the measurements of the bundles derived from the study of the effects found along the walls of Pompeii. Vitruvius offers the correspondence between the diameters of the *modioli* measured in Roman fingers and the weights of the projectiles, expressed in Roman pounds. By knowing the volume of the crushed material, the negative molds of the cavities detected, and the density of the stone material, and, therefore, the relative "breaking strength", it will be possible to reverse engineer the impact dynamics and thereby advance in the interpretation of the marks detected. Assuming that the stone density of Vesuvian basalt and therefore the elasticity coefficient Poisson's ratio of 0.20–0.25 [10], it will be possible to proceed to retrace the criterion formalized by ancient science and proceed further.

The weight of a spherical projectile with a diameter of 140 mm (r = 70 mm) has already been calculated, showing that a slight variation in diameter measurement significantly affects its weight and, therefore, the calibration value of the device.

Sphere volume with radius 70 mm $4/3\pi R^3$ with density 2.5 = 3.6 kg
Sphere volume with radius 80 mm $4/3\pi R^3$ with density 2.5 = 5.5 kg.

Regarding the exact evaluation of the launch tension of the arms of ballistae and scorpions, which influences the initial and final speed of the projectile, it is useful to recall that they were drawn back by a similar loading angle, almost always coinciding with the end of their rotation (90° for the arms oriented in the correct direction—therefore called *eutitone*). Throwing stones of equal weight would have guaranteed identical throws; the fraction of the drawn shot would have been analogous, as is usually used for aimed shots [11].

The same criterion is theoretically applicable to the launch of darts; once the target had been agreed upon, it was necessary to concentrate shots on it using the same launch variables, such as the loading angle of the arms, the length of the dart, and its weight and, obviously, the identical inclination of the scorpion catapult. Both for stone-ball launchers and for dart launchers, aiming devices were used, but were never extremely precise.

Given this calculation, the terminal ballistic parameters are as follows:

- For spheroidal projectiles, based on the formula of Philo of Byzantium [7] (p. 186), the diameter of the hole, measured in fingers, drilled vertically above and below the frame to allow the fiber bundle to pass through, D, is related to the cube root ($\sqrt{}^3$) of the weight of the projectile (a stone carefully polished and standardized in diameter (p) measured in drachmas:

$$D = 1.1 \sqrt{}^3\, p \tag{1}$$

- For the supposed truncated-pyramidal dart tips, which were the most widespread, the calculation is based on the formula reported by Heron (1st century AD). The diameter of the "stretcher" must be proportional to one-ninth of the length of the dart [7] (p. 192).

$$D = L/9 \tag{2}$$

3.1. Spheroidal Traces

For the spheroidal traces (Figure 5) attributed to stones, we have the following:

1. *BallisticData*

Crater diameter Ø 140 mm;
r (radius) = 70 mm; b (ball penetration) = 50 mm;
Normal penetration: 120 mm;
Sphere volume: $V = 4/3\ \pi r^3$; for r = 70 mm, V = 1436 cm^3;
Lava stone density: 2.8 g/cm^3; (3)
Weight of the spherical stone for r = 70 mm, $p \approx 4.0$ kg;
Distance from initial point: 100 m;
Initial velocity = X_0;
Residual impact velocity = X_1.

2. *Diameter of the bundle*

Philo's formula for a spherical projectile: **D = 1.1 $\sqrt{}^3$ p;**
D (diameter in Roman fingers): 1 Roman finger $\approx$ 19 mm;
p (weight of the stone-ball in Attic drachmas): 1 drachma $\approx$ 6 g;
conversion of the weight from kg to drachmas: 4 kg $\approx$ 666.6 drachmas, from which
D = 1.1 $\sqrt{}^3$ 666; D = 1.1 $\times$ 8.73; D $\approx$ 9.6 Roman fingers;
conversion of the diameter from Roman fingers to millimeters;
D = 9.6 fingers $\approx$ 182 mm;
Diameter of the bundle = module $\approx$ Ø 180 mm. (4)

3. *Impact velocity*

E (kinetic energy) = **L** (deformation work); L = **F**orce * displacement, from which
L = **GR** (estimated breaking stress) * **A** (area of the spherical impact crater) $\times$ b (displacement); L = **GR** * **V** (volume of crushed material) becomes
L= π r^2 b + 1/2 * 4/3 πr^3 = 1488 cm^3.
Setting **GR** = 100 kg /cm^2 $\approx$ 1000 N/cm^2, the work performed by the impact force at its end will be
L = 1000 N/cm^2 * 1488 cm^3 = 14,880 N*m.
If all the kinetic energy possessed by the spherical stone is transformed into deformation work of the impact block, we have
E = 1/2 m v^2 with
m (mass in kg) and v (speed in m/s), from which 14,880 N*m.
Assuming that the weight of the spheroidal stone is 4 kg and that the work is equal to the kinetic energy, we have
L = E = 1/2 * 4.0 kg v^2 (kg*m^2/s^2). Solving with respect to v,
v^2 = 14,880 (m^2/s^2)
v = 86.3 m/s. (5)

In the analysis of the motion of the projectile in the air, other types of aerodynamic resistance are not considered.

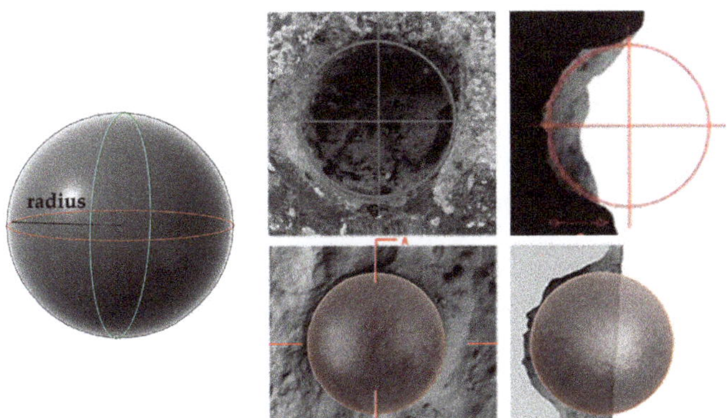

(a) (b)

Figure 5. Spheroidal impact cavity (caused by stone balls): (**a**) Theoretical sphere. (**b**): above, the radius impact measured in 2002; below, renderings of the high-resolution digital models acquired in

2024 [8]. The negative mold provides information on the existing geometry from which the depth and diameter of the theoretical model are derived. The theoretical sphere is derived using automatic region-detected techniques applied to the digital model.

3.2. Quadrangular-Shaped Traces

For the pyramid-shaped traces (Figure 6) caused by iron bolts, we have as follows:

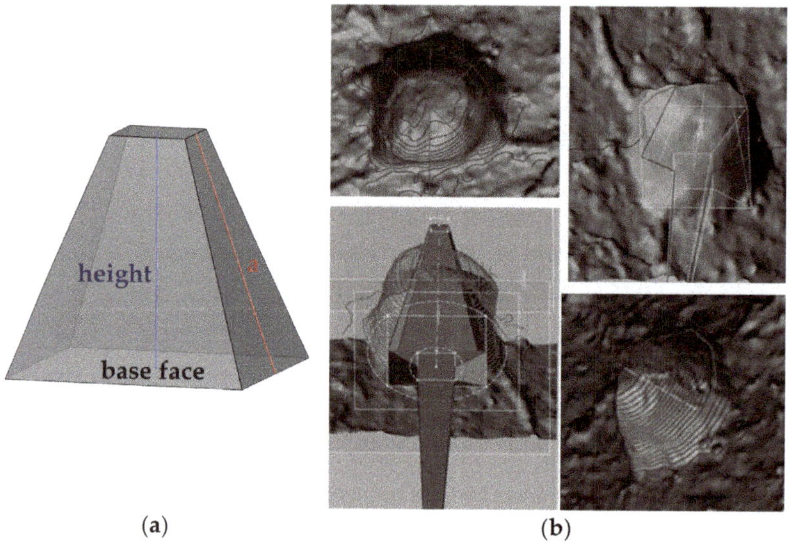

Figure 6. Pyramidal impact cavities (metal tips): (**a**) Theoretical model Pyramidal frustum. (**b**) High-resolution digital models of the small-diameter cavities acquired in February 2024 [12]. The geometric study of the cavities and casts made it possible to assess the best methodological approach to plausibly reconstruct the metal tips of the impact darts. The software tools allow the extraction of best-fitting square sections from specific portions of the mesh and determine the hole's edge length to be approximately 2.5 cm (equivalent to 1/12 of a *pes*, given 1 *pes* = 29.581 cm [13]).

4. *Ballistic Data*

Crater side: 30 mm;
Normal penetration: 30 mm, unpenetrated part 10 mm square pyramid;
Volume $\mathbf{V} = l^2 \times h/3$;
Tip volume $l^2 \times h/3 = 3 \times 3 \times 4/3$ equal to 12 cm^3;
Iron density: 7.8 g/cm^3;
Weight of the dart tip: 93 g;
Weight of the entire dart $\approx$ 150 g;
Distance from the initial point: 100 m
Initial velocity = X_0;
Residual impact velocity = X_1. (6)

5. *Diameter of the Bundle*

Heron's formula: D = 1/9 L;
D (diameter in Roman fingers), 1 finger $\approx$ 19 mm;
Length of the Philo's dart: 481 mm;
Diameter of the scorpion bundle: D = 7.7 module. (7)

6. *Impact velocity*

E (kinetic energy) = L (deformation work); L = Force × displacement.
From which
L = GR (estimated breaking stress) * A (area of the impact crater of the dart) * the displacement; L = GR × V (volume of crushed material) becomes the following:
Setting: GR = 100 $kg/cm^2 \approx 1000\ N/cm^2$;
V (pyramid frustum) = $l^2\ h/3 = 9\ cm^3$;
L = 1000 $N/cm^2 * 9\ cm^3$ = 90,000 N*cm = 900 N*m. (8)
If all the kinetic energy possessed by the dart is transformed into work of deformation of the impact block, we have the following:
E = $1/2\ m\ v^2$ with
m (mass in kg) and v (speed in m/s).
Assuming that the weight of the dart is 0.15 kg and that the work is equal to the kinetic energy, we have
L = E = 900 N*m = $1/2 * 0.15\ v^2$ ($kg*m^2/s^2$). Solving with respect to v, we have
$v^2 = 900/0.5/0.15$ =12,000 (m^2/s^2):
v = 109 m/s

When analyzing the motion of the projectile in the air, other types of aerodynamic resistance are not considered.

Comparing the data, the difference in work (material breakage) caused by iron tips, and rarely bronze, compared to the damage from spherical stone projectiles is evident. The pressure is halved when the impact area is doubled. The approach should therefore be optimized by analyzing the diversity in detail. However, we do not have indications that can provide us with reliable criteria and methods in the light of ancient science. Furthermore, the calibrations described above ignore the presence and significant thickness of the plaster, present at the time of the siege according to some scholars, and of a non-negligible thickness since the Roman plaster itself is several centimeters thick and in this specific case used to hide the fillings resulting from the construction of the towers. The depth detected should therefore be increased by an estimate that, with the data acquired to date, appears arbitrary depending on the thickness and consistency of the mixture with lime and pozzolana. Post-siege evidence is the existence of plaster along the "*murum et plumam*" [14] (p. 151), the fake rustication painted in the background of the portrayed "Fight between Pompeians and Nucerians," which took place in 59 AD (Figure 7) [15].

This criterion, if verified through advanced surveying and reverse-engineering techniques, will allow both virtual and real prototypes to be built in the near future. The module of the machines will be calibrated to the ballistic effects attributed to 1st-century BC artillery and scientifically verifiable. This procedure will distinguish it from other existing prototypes. In fact, various machines have been reconstructed on the basis of the critical interpretation of Greek texts of the *module* of the ballistae by Vitruvius in Book X of *De Architectura*. Reliable reconstructions have been obtained by applying the criteria and principles based on archaeological findings.

Figure 7. Conflict in the amphitheater (or between Pompeians and Nucerians), found in 1869 in the peristyle of the House of the Brawl in the Amphitheatre, I 3, 23 (170 × 185 cm), Pompeii. Conserved in the MANN—National Archaeological Museum in Naples (inv. no. 112222), 1st century AD (59–79 AD). Below: Details (**a**,**b**) of the fresco showing the city walls and towers near the amphitheater. Credits: MANN, https://mann-napoli.it/affreschi/#gallery-13 (accessed on 10 December 2024). (**a**,**b**) Details.

4. Discussion

The spherical impact craters, as widely acknowledged, testify to the use of stone balls of 120–140 mm in diameter. Inside the city, during its excavations, however, numerous rounded stone balls were found, measuring about 270 mm in diameter, a size which corresponds to a weight of about 26 kg, a value coinciding with the ancient unit of weight called "talent" (equal to 26 kg) in the projectiles already mentioned in the historical accounts by Flavius Josephus. At least two questions follow: (i) why were only 140 mm balls used to hit the walls while 270 mm balls were used to hit the inside of the city? (ii) How was it possible for the 270 mm balls to reach over the walls knowing full well that the ballistae barely reached a trajectory grazing the top of the walls resulting in weak and oblique impacts, of little destructive effectiveness, as some of the clearly visible traces testify?

Regarding the first question, it must be assumed that, due to the lower weight of the corresponding ball, its initial and residual speed were higher, with greater devastating effects, also due to the higher pressure that such a ball produced on the planks upon impact, thus requiring a shorter loading time for the relative ballistae. However, there was a second type of elastic artillery: although the criterion informing the elastic torsion machines was, in fact, the same, the performance and structural architecture of the main throwing machines were not. No image supports the existence of a singular stone thrower with a very parabolic shot, remembered by the Greeks as *monoancon* (single arm) and by the Romans as *onager*. However, there is no lack of mentions and traces of its existence, starting with Philo of Byzantium [16] (p. 91–10).

Even Apollodorus of Damascus, who lived in the 1st century AD, Trajan's long-time military engineer, alludes to the *monoancon* [17], and an equally laconic and incomplete mention is found in the histories of Ammianus Marcellinus, who lived in the 4th century AD. He was also substantially contemporary with the Flavius Vegetius Renatus, who lived between the 4th and 5th century AD, who mentioned it thus: "For the same purpose, they make very large stones weighing a talent [stone ball of 26 kg, approximately Ø 270 mm] ... are launched by means of *monoancon*" [18] (IV, chap. 22). From what has been cited, one should conclude that *monoancon* machines did not exist before the very laconic mention by Philo of Byzantium, that is, before the 3rd century BC. But a strange device used in gymnastic races, the *hysplex* (ὑσπληξ), belies this conclusion. The device in question was invented, or rather, co-opted, by Greek engineers to frustrate the asynchronous start of athletes engaged in running races [19].

It appeared in 344–343 BC as attested by a contemporary vase painting, the architectural remains of three stadia located in the north-eastern Peloponnese and, above all, by the discovery of two Hellenistic inscriptions from Delos. This evidence, combined with the hypothesis that the operating mechanism of such a device could have been influenced by the recent invention of the onager, allowed its reconstruction. The *hysplex* consisted of a pair of ropes stretched horizontally between two vertical arms, to which they were tied, respectively, in front of the waist and knees of the runners. The bases of the arms were instead secured in a coil of twisted rope, so that by releasing the restraints the arms would spring towards the ground in a flash, allowing the athletes to start. The device, whose close resemblance to the motion and the conception of the *monoancon*, attests to its existence a few years before the middle of the 4th century BC, perhaps at the court of Philip II of Macedonia. The machine, moreover, was much simpler to build than the ballistae, as there was no need to balance the arms.

But from a ballistic point of view, what was the advantage of this machine in sieges? The ball thrown by its single arm followed a very high parabolic trajectory, with a vertex placed at a height of over 50 m, compared to the plane of the machine's position. Therefore, as the heavy ball ascended, it lost speed and once it reached the vertex its ascending speed was zero. From that moment, it began to fall with a progressive increase in speed, which, therefore, neglecting the air resistance on impact with the ground, had the same initial speed, and was therefore able to penetrate through roofs thanks to its weight. The effect, especially during night-time shooting and with incendiary projectiles, was terrifying, hitting civilian homes. This is a hypothesis that deserves to be properly discussed.

5. Conclusions

The procedure supporting the calculated data is based on the mathematical description of damage and the knowledge handed down from Greek and Roman treatises. The results were partially tested using reconstructions produced by Flavio Russo's *Archeotecnica* for public and private institutions (Figure 8). If the criterion is properly validated, using advanced detection techniques and reverse engineering processes, it will be possible, in the near future, to build (im)material twins [20] of "certified" machines. In fact, the process of quantifying the terminal ballistic parameters makes it possible to determine the module diameter, on the basis of which the weapon's morphology will be proportionate and thus, inductively, the machine's power can be inferred.

(a)

Figure 8. *Cont.*

(b)

Figure 8. Saepinum Archaeological Area, Altilia (CB). Artillery reconstructed by Flavio Russo according to Vitruvius's method (*De Architectura*, Book X). (**a**) Ballista reconstruction, based on the remains found at Hatra (Iraq) and its proportions. (**b**) Catapult (scorpion) reconstruction, based on the remains found at Ampurias and its proportions, compared with Schramm's and Philo of Byzantium's interpretation of the text.

Author Contributions: Conceptualization, F.R. and A.R.; methodology, F.R. and A.R.; software, A.R.; validation, A.R.; formal analysis, F.R. and A.R.; investigation, A.R.; resources, A.R.; data curation, A.R.; writing—original draft preparation, A.R.; writing—review and editing, A.R.; visualization, A.R.; supervision, A.R.; project administration, A.R.; funding acquisition, A.R. All authors have read and agreed to the published version of the manuscript.

Funding: This research received no external funding.

Institutional Review Board Statement: Not applicable.

Informed Consent Statement: Informed consent was obtained from all subjects involved in the study.

Data Availability Statement: The data presented in this study are available in Refs. [21,22].

Acknowledgments: The authors would like to express their gratitude to the management and the appointed officials of the offices of the Pompeii Archaeological Park for granting authorizations for site access and survey operations (MIBACT-SSBA-PES PROTO-ARCH 0006225 14/04/2016 CI. 28.13.07/6).

Conflicts of Interest: The authors declare no conflicts of interest.

References

1. Maiuri, A. *Studi e Ricerche sulla Fortificazione di Pompei*; Monumenti Antichi Pubblicati per Cura Della R. Accademia dei Lincei; Hoepli: Milan, Italy, 1929; Volume 33.
2. Anniboletti, L. Le fasi delle fortificazioni di Pompei. Stato della conoscenza. *SIRIS Studi Ric. Sc. Spec. Beni Archeol. Matera* **2016**, *2015*, 49–70. [CrossRef]
3. Giuseppe, F. *La Guerra Giudaica*; Vitucci, G., Ed.; Mondadori: Milan, Italy, 1989; ISBN 88-04-32627-1.
4. Tucidide. *La Guerra del Peloponneso. Testo Greco a Fronte*; Tosi, R., Ed.; Rusconi Libri: Milan, Italy, 2016; ISBN 978-88-18-03102-7.
5. Maiuri, A. *Introduzione allo Studio di Pompei, a Cura di G. Oscar Onorato, Corso di Antichità Pompeiane ed Ercolanesi 1942-43*; GUF: Naples, Italy, 1943.
6. Van Buren, A.W. Further Studies in Pompeian Archaeology. *MAAR* **1925**, *5*, 103–113. [CrossRef]
7. Gille, B. *Storia delle Tecniche*; Editori Riuniti: Rome, Italy, 1985; ISBN 978-88-359-2903-1.
8. Bertacchi, S.; Gonizzi Barsanti, S.; Rossi, A. Geometry of Wall Degradation: Measuring and Visualising Impact Craters in the Northern Walls of Pompeii. *SCIRES-IT—Sci. Res. Inf. Technol.* **2024**, *14*, 111–128. [CrossRef]
9. Scaglia, S. *Compendio Tecnico al Moderno Tiro di Interdizione*; Centrostampa: Matera, Italy, 2016; ISBN 978-88-940478-9-9.

10. Artini, E. *I Minerali;* Hoepli: Milano, Italy, 1914.
11. Marsden, E.W. *Greek and Roman Artillery. Technical Treatises;* Clarendon Press: Oxford, UK, 1971; ISBN 978-0-19-814269-0.
12. Rossi, A.; Bertacchi, S.; Formicola, C.; Gonizzi Barsanti, S. Piccole indentazioni antropiche rinvenute nella riesumata cinta urbica di Cornelia Veneria Pompeianorum | The small anthropic traces found in the unearthed city walls of Cornelia Veneria Pompeianorum. *Disegnare Idee Immagin.* **2024**, *69*, 54–67. [CrossRef]
13. Lazzarini, M. Metrologia romana. In *Conimbriga;* Universidade de Coimbra: Coimbra, Portugal, 1965; Volume IV.
14. Marcellino, A. *Le Storie;* Selem, A., Ed.; UTET: Turin, Italy, 1976; ISBN 978-88-02-02139-3.
15. De Bernardi, M. La rissa del 59 d.C. nell'anfiteatro pompeiano alla luce di un nuovo ritrovamento archeologico. *Ratio Iuris* **2019**, *XLXVIII*, 1–30.
16. Garlan, Y. Recherches de poliorcétique grecque. In *Bibliothèque des Écoles Françaises d'Athènes et de Rome;* Persée: Lyon, France, 1974; Volume 223.
17. Apollodorus of Damascus Poliorketika. *Poliorkētika kai Poliorkiai Diaphorōn Poleōn | Poliorcétique des Grecs: Traités Thoriques, Récits Historiques;* Wescher, C., Petetin, A., Eds.; Imprimerie Impériale: Paris, France, 1867.
18. Vegezio Publio Flavio Renato. *L'arte della Guerra Romana;* Formisano, M., Ed.; Rizzoli BUR: Milan, Italy, 2003; ISBN 978-88-17-10645-0.
19. Valavanis, P. *Hysplex: The Starting Mechanism in Ancient Stadia. A Contribution to Ancient Greek Technology;* Classical Studies; University of California Press: Berkeley, CA, USA, 1999; ISBN 978-0-520-91592-3.
20. Grieves, M. *Virtually Perfect: Driving Innovative and Lean Products through Product Lifecycle Management;* Space Coast Press: Cocoa Beach, FL, USA, 2011; ISBN 0-9821380-0-8.
21. Russo, F. *L'artiglieria Delle Legioni Romane;* Ist. Poligrafico dello Stato: Roma, Italy, 2004; pp. 30–37, ISBN 88-240-3444-6.
22. Rossi, A. The Survey of the Ballistic Imprints for a Renewed Image of Unearthed Pompeii. *Nexus. Netw. J.* **2024**, *26*, 307–324. [CrossRef]

engineering proceedings

MDPI

Proceeding Paper

Primitive Shape Fitting of Stone Projectiles in Siege Weapons: Geometric Analysis of Roman Artillery Ammunition †

Silvia Bertacchi

Department of Engineering, Università degli Studi della Campania Luigi Vanvitelli, 81031 Aversa, Italy; silvia.bertacchi@unicampania.it

† Presented at the Conference "Discovering Pompeii: From Effects to Causes—From Surveying to the Reconstructions of Ballistae and Scorpiones", Aversa, Italy, 27 February 2025.

Abstract: This paper presents the documentation, study activities, and possible applications of 3D digital models for the analysis and reconstruction of some examples of spheroidal stone projectiles—launched during the Sullan siege in 89 BC—now preserved in the Archaeological Park of Pompeii. The research proposes a methodology to derive best-fitting shapes that most closely adhere to the partially reconstructed image-based geometries. This allows a comparison with the circular ballistic impact traces still present on the ashlars of the northern city walls, as discovered by archaeologists about a hundred years ago. The results facilitate more precise ballistic calculations for the reconstruction of the elastic torsion weapons and their launching power.

Keywords: 3D digital models; ellipsoid fitting; geometric analysis; mesh optimization; photogrammetric survey; Pompeii; reverse modeling; Roman artillery; segmentation; stone missiles

1. Introduction

Integrated digital documentation is currently a widely adopted system in the field of Cultural Heritage (CH) for acquiring reality-based data and using them for the study, analysis, and preservation of the morphometric characteristics of numerous significant architectural and archaeological assets, especially in Italy [1]. In this field, various active and passive devices and sensors, including photogrammetric techniques, have been employed and integrated in recent decades to capture reality as it exists at the moment of acquisition [2–4]—data that can potentially be updated over time through remote monitoring to assess their state of conservation [5,6] while using systems that prioritize non-contact methodologies. This approach is particularly crucial, as surfaces are typically protected and subject to conservation restrictions, permitting only non-invasive analysis methods.

However, digital reality-based models present a major challenge, since, in most cases, the amount of data acquired is neither portable nor easily manageable and requires specific pre-acquisition solutions to limit file size without compromising data quality [7] or the use of online resources to manage very large datasets on hardware with limited local resources [8]. In any case, advanced use remains restricted to specialized skills, preventing experts in other fields from fully exploiting their potential. Beyond the hope for multidisciplinary collaboration, and considering the recent progress in software that has made both programs and online interfaces for browsing 3D material much more user-friendly and accessible [9], the complexity of leveraging more or less structured digital assets, which are useful primarily as metric documentation at a given time, remains a common issue. On the contrary, 3D digital models are extremely useful for studying the original design, for example, providing for a basis for new interpretive perspectives [10], enabling digital reconstructions, and supporting virtual anastylosis [11]. They can also undergo manual or automatic segmentation processes and data extraction for interpretive purposes or for advanced study of their constitutive geometry, even when elements are incomplete [12].

Within the framework of the MUR-PRIN 2022 SCORPiò-NIDI project, and as a complement to the study of the ballistic imprints found on the northern city walls of Pompeii, originating from traces left during the Sullan siege (1st century BC) [13,14], this paper presents the documentation and analysis of the spheroidal stone projectiles (Figure 1) that most likely produced the marks on the ashlars.

Figure 1. Several stone projectiles attributed to the siege of Sulla in the 1st century BC are preserved in the Antiquarium of Porta Marina. Scholars have also recorded additional ones kept in the excavation deposits [15] (pp. 33, 93), but the initial study focuses only on the stone projectiles displayed to the public, here visible in the former exhibition layout (**a**,**b**).

2. Materials and Methods

This study proposed a methodology for the reality-based documentation, geometric analysis, and theoretical reconstruction of spheroidal stone projectiles attributed to the Sullan siege of Pompeii (89 BC). The projectiles, launched by Roman elastic–torsion siege weapons such as ballistae and onagers, as described by Vitruvius [16], provide valuable evidence of the throwing power of ancient artillery. Similar findings of Roman catapult ammunition have been reported in various sites [17–19], but the Pompeii specimens are unique because their destructive effects are still visible on the city's northern walls, found between Vesuvio and Ercolano Gate.

Thus, the research aimed to reconstruct the most geometrically reliable shapes and compare them with the spherical impact traces on the ancient city walls, first documented by archaeologists a century ago [20]. Until recently, in a previous exhibition layout, these projectiles were displayed in the Antiquarium, where they were partially documented with Structure from Motion (SfM) techniques to facilitate the geometric analysis and testing of different 3D reconstruction approaches, both manually and automatically. Using an integrated workflow, survey techniques and geometric analysis were applied to these small-sized artifacts, acquired through rapid photogrammetric techniques. Their virtual models—partially reconstructed—served as a basis for theoretical geometric reconstructions for comparison with the ballistic indentations on Pompeii's northern walls.

At the end of this study, the analysis helped to determine whether any stone projectiles matched the dimensions of the impact traces attributed to the Sullan siege, potentially identifying the ammunition responsible for the marks on the ashlars. The research also provided recommendations for further investigations on the topic.

2.1. The Case Study

The stone balls of various sizes and weights used for ballistae, launched by pulling a rope on a rail in Roman elastic–torsion weapons, were approximately spherical in shape. In Pompeii, some of the larger and heavier balls have been found mainly in the northwestern part of the city (Regio VI, e.g., the House of the Labyrinth), where the most intense clashes took place, as well as in several private domus near the most affected areas [21]. Scholars believe that the larger projectiles, such as stone balls weighing one Greek talent (equivalent to about 26 kg), were thrown over the walls using ballistae or catapults, capable of hurling them beyond 200 m [22], or onagers, which launched them in parabolic trajectories to overcome the fortified walls present at the time.

Regarding the projectiles of different calibers, materials, and hardnesses that were collected in large numbers during the siege of the city, in particular the stone balls likely used by Sulla's forces in 89 BC, the archaeologist Albert W. Van Buren already mentioned them in the 1930s. In his writings, he recorded 86 specimens "of the type of ballista ball" [23] (p. 15) (Figure 2) preserved in the brick building along the northern part of the western side of the Forum. In all probability, they represented a selection of the surviving ballistic projectiles thrown by the besieging Sullan army against the walls of Pompeii, along with a portion of the projectiles thrown by the city's defenders. Such defensive missiles were usually stored near the fortified structures (e.g., those depicted by Piranesi in the nineteenth century near the Ercolano Gate [24] (vol. 1, Plate 5)). However, after the war emergency, they were often relocated or possibly repurposed for other uses.

Figure 2. A comparison of (left) the presumed blunt projectile launched and (right) its presumed effect on the city walls (not to scale). Historical pictures: (**a**) "Missiles in the Museum of the Forum" (Plate 2, Figure 1, 1932); (**b**) "Marks of the Sullan Bombardment, Pompeii" (Plate 60, Figure 1, 1925). Photograph credits: [23,25].

Of the stone balls photographed at that time, a selection had been made by the archaeologist himself who, with the approval of the Superintendence, ensured the removal of certain stones, that, in his opinion, did not correspond to the type of "missile" used for launching.

The selected collection of stones was undoubtedly quite distinct from both those used as weights (see Figure 2a, above right) and from irregular yet rounded stones made of various materials that must have served for other purposes. The projectiles were well rounded rather than irregular, primarily made of tuff, with some examples of lava, sandstone, and limestone (but none of marble). They lacked no intentional inscriptions or markings, but many exhibited damaged areas on parts of their surfaces, possibly due to impact or partial preservation. Approximate recorded diameters for typical specimens ranged from 10 to 23 cm. More specifically, Van Buren [23] (p. 16) lists examples measuring 0.23, 0.21, 0.19, 0.16, 0.14, 0.13, 0.11, and 0.10 m, without mentioning any conversion to

Roman units of measurement. In his view, these were comparable to the 2500 examples found at the military port of Carthage, which ranged from 10 to 30 cm in diameter (p. 16, note 2).

In any case, the projectiles recovered from different parts of the excavation site lacked any documentation regarding the original location of individual pieces. Despite the archaeologist's efforts to trace records of their discovery in excavation reports from his time [26,27], no definitive conclusions could be drawn about their primary context or the locations from which they were fired. Furthermore, he himself did not provide precise measurements or weights for the specimens, leaving such details to be determined in future, with more detailed studies. He only referenced similar missiles found at various archaeological sites in the 1920s–30s, comparable to the Pompeian examples, and discovered at other key sites besieged by the Romans [28].

The discovery of 215 lead sling bullets and 17 stone ballista balls—some partially fragmented and made of either limestone or dark grey lava stone with black inclusions—within three different calibers, averaging 18, 21, and 25 cm in diameter, was a more recent find. These were unearthed during the Anglo-American Project at Pompeii, during excavations conducted between 1996 and 2003 near the Ercolano Gate in Regio VI, Insula I [29] (p. 1).

2.2. Three-Dimensional Documentation

For the initial analysis, a selection of the approximately 50 stone balls were documented using rapid photogrammetry for 3D reconstruction (Figure 3). Despite challenging lighting conditions and the stacked arrangement of the stone balls in the former exhibition layout, it was possible to reconstruct 3D models of only the visible portions of some of the objects. However, at this stage of the research, the hidden parts remain undocumented; the objects should be relocated to a more suitable environment for complete documentation.

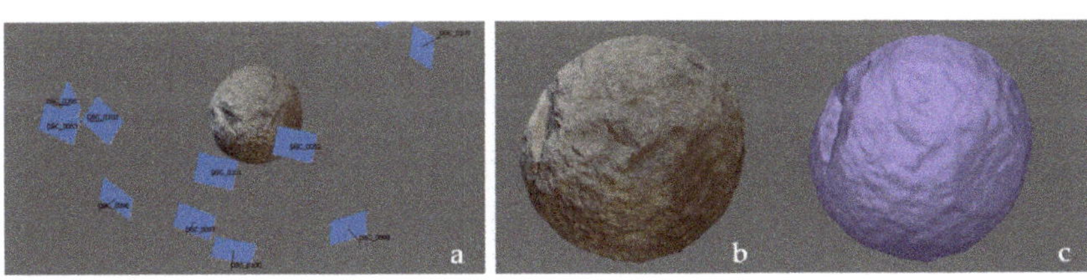

Figure 3. Image processing phase with Metashape (**a**); reconstructed polygonal mesh model, shown with (**b**) and without texture (**c**).

The acquired images (63 photos, 6000 × 4000 pixel, NIKON D5200, focal length 18 mm, ISO sensitivity 450, exposure time 1/60) were processed with Agisoft Metashape Professional software (version 2.1.1) [30] to generate a geometric reconstruction of the pile of stone balls in their fixed position (as their relocation was not possible). The software also generated a textured model representing the apparent color, resulting in partial reconstruction. However, the full study is in progress, various models were evaluated in this paper, and only the most complete ones were selected for the following steps, i.e., those with a reconstructed surface exceeding 50%, while also considering objects of different sizes (large diameter: >16 cm; medium: 12–16 cm). Meanwhile, elongated ovoid objects, which seemed to belong to a different category, were excluded. The resulting partial models were then exported for post-processing using Reverse Modeling software 3D Systems Geomagic Design X (version 2016.1.1) [31] to perform dimensional and geometric analyses. The models featured a polygonal mesh with an average edge length of 1.2 mm, which was doubled after optimizations.

Beyond the operational steps, the reconstruction of the existing projectile geometry (12–18 cm in diameter) required addressing key methodological aspects for defining a reliable theoretical shape and determining the center and radius of the ideal sphere closest to the real object. The main challenges included the following: (i) some reconstructed portions may not have accurately reflected reality, especially in occluded or poorly lit areas; (ii) since the initial survey was not metric, scaling errors may have occurred, potentially affecting the final results.

Therefore, future documentation with active sensors is planned; (iii) theoretical shapes will have limited adherence to stone projectiles, which are typically rough and irregular, with both chipped areas and tool marks. Thus, the analysis will focus on defining ranges of spheres that encompass the average or total surface area; (iv) objects with a low percentage of reconstructed surface are inherently less reliable than more complete models.

For this reason, the main processing steps performed in this research were as follows (Figure 4):

1. Initial mesh editing: The lower, less documented, and more uncertain part was removed by trimming the edges to eliminate unreliable polygonal surfaces. Most image processing software also includes built-in tools for filtering and assessing model confidence (a). Irrelevant parts could be manually removed, especially because the objects were stacked, leading to unintended geometry reconstructions. The boundary editing tool extended the selection to refine edge regions affected by poor lighting or contact with other objects. Finally, an edge smoothing process was applied to improve surface continuity (b).
2. Surface refinement: Smoothing operations helped to eliminate inconsistencies caused by the presence of moss, lichen, or non-removable weed vegetation.
3. Mesh cleaning: Small hole-filling operations (within a certain size) and automatic corrections (e.g., removal of self-intersecting faces, minor and pendant faces, and clusters) improved the model's accuracy (c);
4. Mesh optimization: Additional refinements were applied to regularize the mesh structure (d).

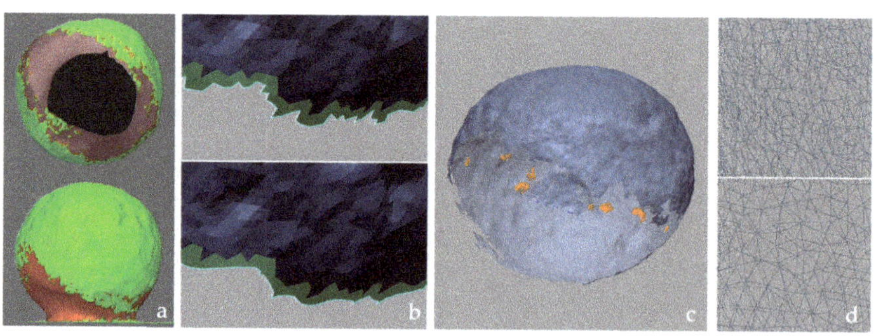

Figure 4. Processing steps for mesh editing: the model confidence of the reconstructed mesh (**a**); edge smoothing before and after application (**b**, above-bottom); mesh editing and cleaning (**c**); mesh optimization before and after application (**d**, above-bottom).

Once edited, the optimized models were re-textured into the photogrammetry software by reloading them with their original coordinates but with improved geometry. Due to the close stacking of objects, reprojection errors may have affected unrelated areas of the 3D model. These could be corrected by applying masks before generating the final texture.

3. Results

The geometric analysis of the obtained meshes can be conducted through different approaches: (i) Traditional methods, which involve taking approximate measurements and attaching photographs, as carried out in past studies [23,25,32]. While this provides valuable general information, it lacks precision. (ii) Manual Surface Reconstruction (MSR), where an expert operator extracts significant sections from the real survey to reconstruct the most accurate 3D shapes. This approach offers greater adherence to reality but is time-consuming. (iii) Automatic Surface Extraction (ASE), using software tools that approximate the objects as perfect theoretical spheres.

For MSR (Figure 5), a theoretical surface can be manually reconstructed by identifying key sections for shape analysis. This involves selecting multiple planes that intersect the existing portion of the object's mesh at predefined intervals (a). The resulting polylines reliably describe the surface's neighborhood and help to identify the most relevant sections for further modeling. From these polylines, best-fitting (B-F) circles can be obtained by excluding segments belonging to depressions (highlighted in pink), which result from material loss due to impact or natural stone defects during manufacturing (b). Alternatively, for a more precise reconstruction using loft surfaces, connecting different profiles (c), a set of B-F curves (ellipses) can be derived (d).

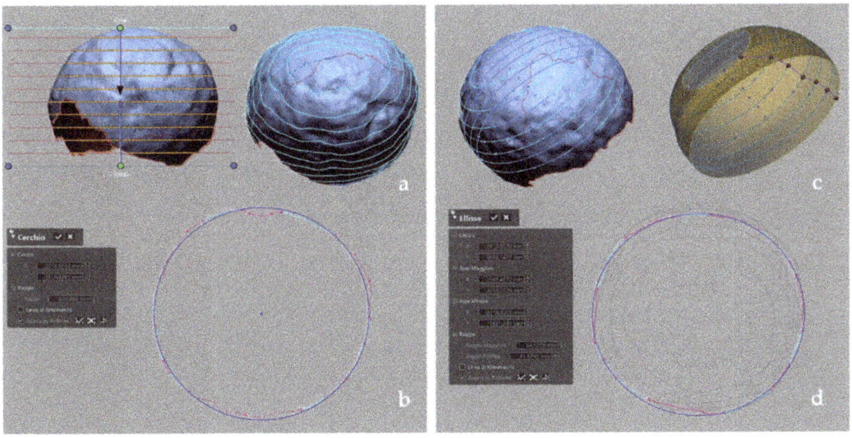

Figure 5. Processing steps for MSR reconstruction: (**a**) multiple plane section for B–F main circle extraction (**b**); (**c**,**d**) B-F ellipses extraction for loft reconstruction.

For ASE (Figure 6), using automatic primitives, it is possible to quickly extract simple solid geometric objects from a regionalized mesh or a portion of it. After filtering, the model must be automatically segmented, applying settings that account for mesh smoothness and uniformity, adjustable by the operator (a). This process segments the model into distinct color-coded regions, which the software automatically classifies as either freeform shapes or approximations of standard solids (spheres, cylinders, etc.). These regions, or a selected subset, such as excluding hollow parts caused by material breakage, can be used to extract surfaces or solids conforming to basic geometric primitives. In this case, the extracted shape is explicitly defined as a (partial or complete) sphere (b). The Automatic Surface Extraction suggests a spherical surface representing the average of the considered regions. However, it does not fully encompass the actual geometry; instead, it generates an approximate theoretical sphere. At this stage, the operator can adjust various parameters of the generated shape, including the radius and position of the semicircular profile of revolution, the position of the sphere's center, and the vector passing through the center, acting as the axis for the 360° revolution of the profile. For instance, increasing the dimensions to contain the entire geometry requires modifying the semicircular profile's radius. This adjustment affects the final reconstructed surface size (c). This approach is useful because initial analyses have shown that the geometry of the examined projectiles does not correspond to a perfect sphere but rather to a spheroid (d). Extracting multiple sections from the mesh reveals variations in radius measurements (e), sometimes differing by a few millimeters across different sections of the irregular mesh. By considering a set of three mutually orthogonal reference planes, it is necessary to define, especially for virtual simulations, whether to adopt a maximum or minimum bounding sphere within the given range. To visually verify deviations between the mesh and the reconstructed spherical surface, the Accuracy Analyzer tool can be used. This function measures deviations between two mesh bodies (base entity vs. reference entity), displaying results through a color-coded deviation bar. The scale ranges from yellow to red for positive values and light blue to dark blue for negative values. Additionally, the operator can set a tolerance threshold and analyze a histogram to monitor the distribution of deviation values (f).

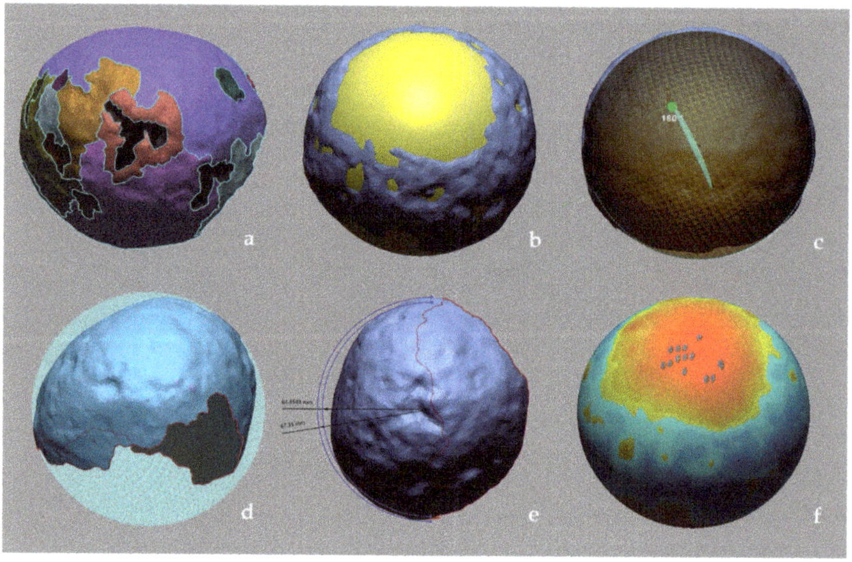

Figure 6. Processing steps for ASE reconstruction: (**a**) mesh regionalization; (**b**) automatic sphere extraction; (**c**) parameter modification; (**d**) spheroid case study; (**e**) minimum/maximum radius for the bounding sphere; (**f**) deviation between mesh and theoretical sphere.

4. Discussion and Future Developments

Regarding shape completion, minimizing manual user intervention, traditionally employed for completing missing parts, as discussed in previous sections, can be achieved by transitioning toward automated systems based on point clouds and machine learning [33]. Recent advances in single-frame shape reconstruction, including organic forms [34], could also be explored. However, these approaches require validation when dealing with objects originally designed to have specific geometric shapes, as our study focuses on the geometric analysis of ancient artifacts.

Although the methods used provided acceptable solutions, they proved less suitable when the object's morphology was ovoidal and deviated significantly from a theoretical sphere, resembling instead an ellipsoid (Figure 7).

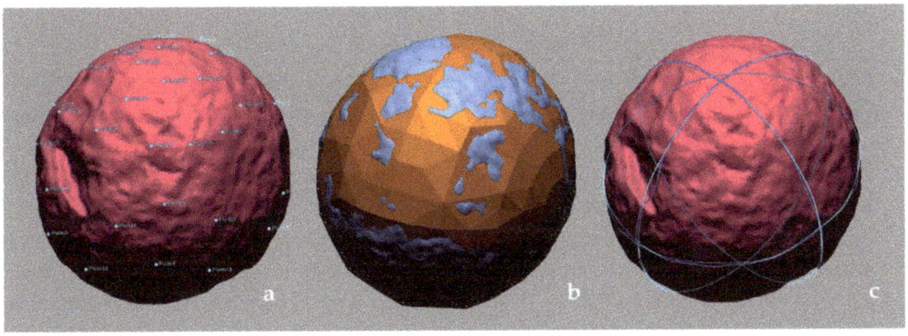

Figure 7. Processing steps for automatic ellipsoid reconstruction: (**a**) selected surface points; (**b**) automatically generated 3D ellipsoid using the LSEF method implemented in ChatGPT; (**c**) cross-sections deviation analysis between mesh and 3D ellipsoid.

Rather than relying on automatic shape prediction, achieving a more accurate reconstruction would require generating a B-F ellipsoid directly from the available polygonal mesh [35]. This process requires a sufficiently large and evenly distributed set of surface points, ensuring that they are not located in depressions, fractures, or uncertainly reconstructed areas, to reliably approximate the object's shape. To create a 3D ellipsoid from a given set of points, a fitting algorithm such as the Least Squares Ellipsoid Fitting (LSEF) method is typically used [36]. This method, based on least squares minimization, calculates the ellipsoid parameters (center, principal axes, and orientation) to best match the data. A test case was, therefore, performed using 30 surface points (a). The best-fitting ellipsoid was computed and found to closely align with the object's shape. The calculation can be manually performed using Python 3.13.4 [37] or Rhinoceros/Grasshopper [38,39], processing spatial coordinates extracted from selected surface points. However, for this case study, the reconstruction was conducted automatically, providing only the spatial coordinates. The LSEF method, implemented via ChatGPT-4 [40], calculated key parameters (semi-axes, center position, orientation, and shape) and determined the best-fitting ellipsoid for the given partial surface (b). The method then proceeded to generate a polygonal mesh, allowing for a deviation analysis (c).

In the case examined, the object did not deviate significantly from a sphere, as the semi-axis values were nearly equal. However, the integration of AI-driven computation significantly reduced processing and data verification times. This method can be applied to future cases for broader and more in-depth studies.

5. Conclusions

The opportunity to document not only the impact craters on the city walls but also the stone projectiles that most likely caused them represents a unique chance to digitally compare both the dimensions of the holes and the blunt objects. This allows us to define a size range for the stone balls involved in the impacts, which is also essential for future virtual simulations of the firing power of Roman artillery.

This initial test, conducted exclusively through photogrammetric survey, will yield more precise dimensional data once a structured-light or triangulation scanner can be used. Such a method would provide not only higher resolution and detail quality but also greater measurement accuracy. Nevertheless, the current survey method already ensures a higher level of measurement reliability than what was achievable in antiquity (1st century BC) when producing and replicating stone projectiles. Therefore, the adopted procedure is expected to yield equally reliable and realistic results.

As a general consideration, once reliable measurements are obtained—currently sufficient for a preliminary study but better refined through range-based acquisition systems for further research on projectile mass—they can be compared to ancient units of measurement to identify potential discrepancies. For instance, the standard length of the Roman palmus was equal to one-quarter of a pes (Roman foot) and measured approximately 7.41 cm [41] (p. 8). This measurement aligns with the diameters of some of the impact cavities analyzed and with stone projectiles measuring 7 or 8 digiti in diameter (Roman digitus, each ≈ 1.85 cm, yielding approximately 12.95 cm or 14.80 cm, respectively), as found in the case studies. The data confirm the classification of two projectile size groups: larger projectiles, with diameters exceeding 9 Roman digiti (≈16.65 cm); smaller projectiles, with diameters ranging between 7 and 8 digiti (≈12–15 cm). The latter dimensions fall within the range proposed by Van Buren for the impact marks observed on the city walls [23] (p. 16), while according to Burns, only the 18 cm diameter balls (weighing 15 mina) matched the impact marks by physically fitting the missile into them [29] (p. 6).

Similar sandstone projectiles used in Roman artillery (ballistae) have been recorded in other countries, showing comparable dimensions with an average diameter ranging between 13.5 and 14.5 cm, with the average value equal to 14.3, corresponding to 9–10 librae balls [18] (Figure 2, p. 70). These findings confirm that beyond their roughly spherical shape, which remained imperfect due to the need for rapid production, the preferred size of projectiles was dictated by the mechanical constraints of the ballistae. The weight of these projectiles was calibrated to match the standard firing power of Roman siege weapons.

Further research and ongoing analysis of a larger sample of digitized case studies will contribute to a deeper understanding of the stone projectiles used in Roman artillery, particularly those probably used during the Sullan siege of Pompeii.

Funding: This research was supported by the project "SCORPiò-NIDI", CUP B53D23022100006 (DD n. 1012/2023), funded by the Italian Ministry of Research under the PRIN (call DD n. 104/2022) funding initiative.

Institutional Review Board Statement: Not applicable.

Informed Consent Statement: Not applicable.

Data Availability Statement: Research data are available upon request.

Acknowledgments: The author would like to express her gratitude to the management and the appointed officials of the offices of the Pompeii Archaeological Park for granting authorizations for site access and survey operations.

Conflicts of Interest: The authors declare no conflicts of interest.

Abbreviations

The following abbreviations are used in this manuscript:

3D	Three-Dimensional
ASE	Automatic Surface Extraction
CH	Cultural Heritage
B-F	Best-fitting
LSEF	Least Squares Ellipsoid Fitting
MSR	Manual Surface Reconstruction
SfM	Structure from Motion

References

1. Fico, D.; Rizzo, D. (Eds.) *Conservation Methodologies and Practices for Built Heritage*; MDPI—Multidisciplinary Digital Publishing Institute: Basel, Switzerland, 2025; ISBN 978-3-7258-2743-5.
2. Guarnieri, A.; Vettore, A.; El-Hakim, S.; Gonzo, L. Digital Photogrammetry and Laser Scanning in Cultural Heritage Survey. In Proceedings of the ISPRS XX Symposium, Istanbul, Turkey, 15 June 2004; Volume 2, pp. 1–6.
3. Fassi, F.; Achille, C.; Fregonese, L.; Monti, C. Multiple Data Source for Survey and Modelling of Very Complex Architecture. *Int. Arch. Photogramm. Remote Sens. Spat. Inf. Sci.* **2010**, *XXXVIII*, 234–239.
4. Fascia, R.; Barbieri, F.; Gaspari, F.; Ioli, F.; Pinto, L. From 3D Survey to Digital Reality of a Complex Architecture: A Digital Workflow for Cultural Heritage Promotion. *Int. Arch. Photogramm. Remote Sens. Spat. Inf. Sci.* **2024**, *XLVIII-2-W4-2024*, 205–212. [CrossRef]
5. Astorga González, E.M.; Municio, E.; Noriega Alemán, M.; Marquez-Barja, J.M. Cultural Heritage and Internet of Things. In *GoodTechs '20: Proceedings of the 6th EAI International Conference on Smart Objects and Technologies for Social Good, Antwerp, Belgium, 14–16 September 2020*; Association for Computing Machinery: New York, NY, USA, 2020; pp. 248–251. [CrossRef]
6. Casillo, M.; Colace, F.; Gaeta, R.; Lorusso, A.; Santaniello, D.; Valentino, C. Revolutionizing Cultural Heritage Preservation: An Innovative IoT-Based Framework for Protecting Historical Buildings. *Evol. Intell.* **2024**, *17*, 3815–3831. [CrossRef]
7. Costantino, C.; Prati, D.; Predari, G.; Bartolomei, C. 3D Laser Scanning Survey for Cultural Heritage. A Flexible Methodology to Optimize Data Collection. *Int. Arch. Photogramm. Remote Sens. Spat. Inf. Sci.* **2020**, *XLIII-B2-2020*, 821–828. [CrossRef]
8. Li, F.; Spettu, F.; Achille, C.; Vassena, G.; Fassi, F. The Role of Web Platforms in Balancing Sustainable Conservation and Development in Large Archaeological Site: The Naxos Case Study. *Int. Arch. Photogramm. Remote Sens. Spat. Inf. Sci.* **2024**, *XLVIII-2-W8-2024*, 303–310. [CrossRef]
9. Spiess, F.; Waltenspül, R.; Schuldt, H. The Sketchfab 3D Creative Commons Collection (S3D3C). *arXiv* **2024**, arXiv:2407.17205. [CrossRef]
10. Bertacchi, S.; Juan Vidal, F.; Fantini, F. Ancient Architectural Design Interpretation: A Framework Based on Alexandrian Manuals. *Acta IMEKO* **2024**, *13*, 1–9. [CrossRef]
11. Adembri, B.; Cipriani, L.; Bertacchi, G. Guidelines for a Digital Reinterpretation of Architectural Restoration Work: Reality-Based Models and Reverse Modelling Techniques Applied to the Architectural Decoration of the Teatro Marittimo, Villa Adriana. *ISPRS-Int. Arch. Photogramm. Remote Sens. Spat. Inf. Sci.* **2017**, *XLII-5/W1*, 599–606. [CrossRef]
12. Eramo, E.; Cinque, G.E. The Vault of the So-Called Serapeum: An Ellipsoidal Geometry at Hadrian's Villa. In *Graphic Horizons*; Hermida González, L., Xavier, J.P., Pernas Alonso, I., Losada Pérez, C., Eds.; Springer Nature Switzerland: Cham, Switzerland, 2024; pp. 51–58. [CrossRef]
13. Rossi, A. The Survey of the Ballistic Imprints for a Renewed Image of Unearthed Pompeii. *Nexus Netw. J.* **2024**, *26*, 307–324. [CrossRef]
14. Bertacchi, S.; Gonizzi Barsanti, S.; Rossi, A. Geometry of Wall Degradation: Measuring and Visualising Impact Craters in the Northern Walls of Pompeii. *SCIRES-IT-Sci. Res. Inf. Technol.* **2024**, *14*, 111–128. [CrossRef]
15. Russo, F.; Russo, F. *89 a.C. Assedio a Pompei: La Dinamica e Le Tecnologie Belliche Della Conquista Sillana Di Pompei*; Edizioni Flavius: Pompei, Italy, 2005; ISBN 88-88419-32-2.
16. Vitruvius, P. *De Architectura*; Gros, P., Ed.; Einaudi: Torino, Italy, 1997.
17. Holley, A.E. The Ballista Balls from Masada. In *Masada IV: The Yigael Yadin Excavations 1963–1965. Final Reports*; Aviram, J., Foerster, G., Netzer, E., Eds.; Israel Exploration Society: Hebrew University of Jerusalem: Jerusalem, Israel, 1994; pp. 347–365. ISBN 978-965-221-010-4.
18. Wilkins, A.; Barnard, H.; Rose, P.J. Roman Artillery Balls from Qasr Ibrim, Egypt. *Sudan Nubia Sudan Archaeol. Res. Soc.* **2006**, *10*, 64–78.
19. Wilkins, A. *Roman Imperial Artillery: Outranging the Enemies of the Empire*, 3rd ed.; Archaeopress Roman Archaeology; Archaeopress: Oxford, UK, 2024; ISBN 978-1-80327-783-7.
20. Maiuri, A. *Studi e Ricerche Sulla Fortificazione di Pompei*; Monumenti Antichi Pubblicati per cura della R. Accademia dei Lincei; Hoepli: Milano, Italy, 1929; Volume 33.
21. Giglio, M.; Pesando, F. La Guerra sociale, l'assedio e la fine di Pompei sannitica. In Proceedings of the Pompei 79 d.C. Una Storia Romana, Roma, Italy, 6 November 2020–31 January 2021; Torelli, M., Ed.; Electa: Milano, Italy, 2020; pp. 144–151, ISBN 978-88-9282-011-1.
22. Di Marco, U.; Molari, P.G. Archimede ed il sistema di caricamento della balista da un talento utilizzata nella fortezza dell'Eurialo di Siracusa. In Proceedings of the 4th International Conference | Atti dell'8° Convegno Nazionale, Naples, Italy, 11 December 2020; D'Agostino, S., d'Ambrosio Alfano, F.R., Eds.; Cuzzolin: Napoli, Italy, 2020; Volume I, pp. 393–408.
23. Van Buren, A.W. Further Pompeian Studies. *MAAR* **1932**, *10*, 7–54. [CrossRef]
24. Piranesi, F.; Guattani, G.A. *Antiquités de la Grande Grèce, Aujourd'hui Royaume de Naples, Gravées par François Piranesi*; Piranesi Leblanc: Paris, France, 1804; Volume 1.

25. Van Buren, A.W. Further Studies in Pompeian Archaeology. *MAAR* **1925**, *5*, 103–113. [CrossRef]
26. Fiorelli, G. *Pompeianarum Antiquitatum Historia*; s.n.: Napoli, Italy, 1862; Volume 2.
27. Fiorelli, G. *Gli Scavi di Pompei dal 1861 al 1872 Relazione al Ministro della Istruzione Pubblica di Giuseppe Fiorelli*; Tipografia italiana nel liceo V. Emanuele: Napoli, Italy, 1873.
28. Schumacher, G. *Tell El Mutesellim. Bericht Über Die 1903 Bis 1905 Mit Unterstützung SR. Majestät Des Deutschen Kaisers Und Der Deutschen Orientgesellschaft Vom Deutschen Verein Zur Erforschung Palästinas Veranstalteten Ausgrabungen*; Rudolf Haupt: Leipzig, Germany, 1908.
29. Burns, M. Pompeii under Siege: A Missile Assemblage from the Social War. *JRMES* **2015**, *14/15*, 1–10.
30. Agisoft Metashape Professional. Available online: https://www.agisoft.com/ (accessed on 1 February 2025).
31. 3D Systems Geomagic Design X. Available online: https://www.3dsystems.com/software/geomagic-design-x (accessed on 1 February 2025).
32. Maiuri, A. Pompei. Isolamento della cinta murale tra la Porta Vesuvio e la Porta Ercolano. *Not. Degli Scavi Antich.* **1943**, *VII*, 275–314.
33. Tesema, K.W.; Hill, L.; Jones, M.W.; Ahmad, M.I.; Tam, G.K.L. Point Cloud Completion: A Survey. *IEEE Trans. Vis. Comput. Graph.* **2024**, *30*, 6880–6899. [CrossRef] [PubMed]
34. Diao, H.; Jiang, X.; Fan, Y.; Li, M.; Wu, H. 3D Face Reconstruction Based on a Single Image: A Review. *IEEE Access* **2024**, *12*, 59450–59473. [CrossRef]
35. Eramo, E.; Fantini, F. An Integrated Approach for Investigating Roman Cupolas: From Segmented Models to Trikentron Analyis. *Disegnarecon* **2024**, *17*, 5.1–5.12. [CrossRef]
36. Li, Q.; Griffiths, J.G. Least Squares Ellipsoid Specific Fitting. In Proceedings of the Geometric Modeling and Processing 2004 (GMP'04), Beijing, China, 13–15 April 2004; pp. 335–340. [CrossRef]
37. Python. Available online: https://www.python.org (accessed on 23 March 2025).
38. Rhinoceros (Rhino 3D). Available online: https://www.rhino3d.com (accessed on 1 February 2025).
39. Grasshopper. Algorithmic Modeling for Rhino. Available online: https://www.grasshopper3d.com (accessed on 23 March 2025).
40. ChatGPT (OpenAI). Available online: https://openai.com/chatgpt (accessed on 2 April 2025).
41. Molari, P.G.; Maraldi, M.; Angelini, G.; Bignami, S.; Lionello, G. La ricostruzione della balista di Vitruvio. In Proceedings of the Quinta Giornata di Studio Ettore Funaioli (V GEF), Bologna, Italy, 15 July 2011; 2012; pp. 1–26. [CrossRef]

engineering proceedings

MDPI

Proceeding Paper

Tracing Metal Dart Impacts Through 3D Reverse Modeling on the Northern Walls of Pompeii †

Adriana Rossi * and Silvia Bertacchi

Department of Engineering, Università degli Studi della Campania Luigi Vanvitelli, 81031 Aversa, Italy; silvia.bertacchi@unicampania.it

* Correspondence: adriana.rossi@unicampania.it

† Presented at the Conference "Discovering Pompeii: From Effects to Causes—From Surveying to the Reconstructions of Ballistae and Scorpiones", Aversa, Italy, 27 February 2025.

Abstract: This study examines the first systematic documentation of a series of small impact marks on the northern walls of Pompeii, interpreted as the result of Roman metal dart projectiles launched during the Sullan siege in 89 BC. Using high-resolution, reality-based 3D models, comparative analysis, and reverse modeling techniques, the research explores the hypothesis that a distinctive fan-shaped configuration of quadrangular indentations may have been produced by a repeating catapult, known as the *polybolos*. The integration of close-range photogrammetry, laser scanning, and digital reconstruction tools demonstrates how virtual casts and comparative modeling can contribute to archaeological interpretations of ancient projectile weaponry.

Keywords: 3D documentation; ballistic impacts; dart launcher; digital archaeology; *polybolos*; Pompeii; reverse modeling; Roman artillery; siege warfare; Sullan siege

1. Introduction

Aware of the exceptional nature of the archaeological site of ancient Pompeii, the survey activities carried out as part of the SCORPIò-NIDI project were aimed to establish a methodology for classifying the numerous indentations visible along the stretch of the city walls between Vesuvio Gate and Ercolano Gate (Figure 1), which can reasonably be assumed to be of anthropic origin [1].

For the circular or spherical cavities, a first framework for classification was originally proposed by archaeologist Albert William Van Buren in the late 1920s [2]. Based on the diameters of the holes in the stone ashlars and their correlation to the size of the rounded stone projectiles found at the site, the archaeologist interpreted the impact marks as the result of stone balls launched during the siege led by Sulla in 89 BC—primarily causing compressive damage upon impact [2,3] (Figure 2a). A more recent study focusing on lead sling bullets of different calibers [4] has documented a concentration of small-scale elongated ovoidal impacts peppering the area near Porta Ercolano (Figure 2b).

However, when it comes to cavities attributed to metallic dart impacts (Figure 2c), there are no specific studies—only general references to indentations left by catapult projectiles. The study of this category of evidence is more complex. Firstly, the physical and chemical degradation of ferrous materials over time either prevented or severely limited the possibility of recovering arrowheads from the Sullan period left on the field or walls. Moreover, the energy transmitted during the impact with the extrados of the wall, made of Nocera tuff or Sarno limestone [5,6], resulted in splintering the stone and surrounding fragments. As a result, the shape of the ballistic mark can differ substantially from the geometric shape of the original metal arrowhead, making its location and interpretation difficult—especially when considering the cavity-rich texture of the wall's stone.

Figure 1. Study area between Vesuvio and Ercolano Gates, Archaeological Park of Pompeii. Source: Google Earth, year 2023.

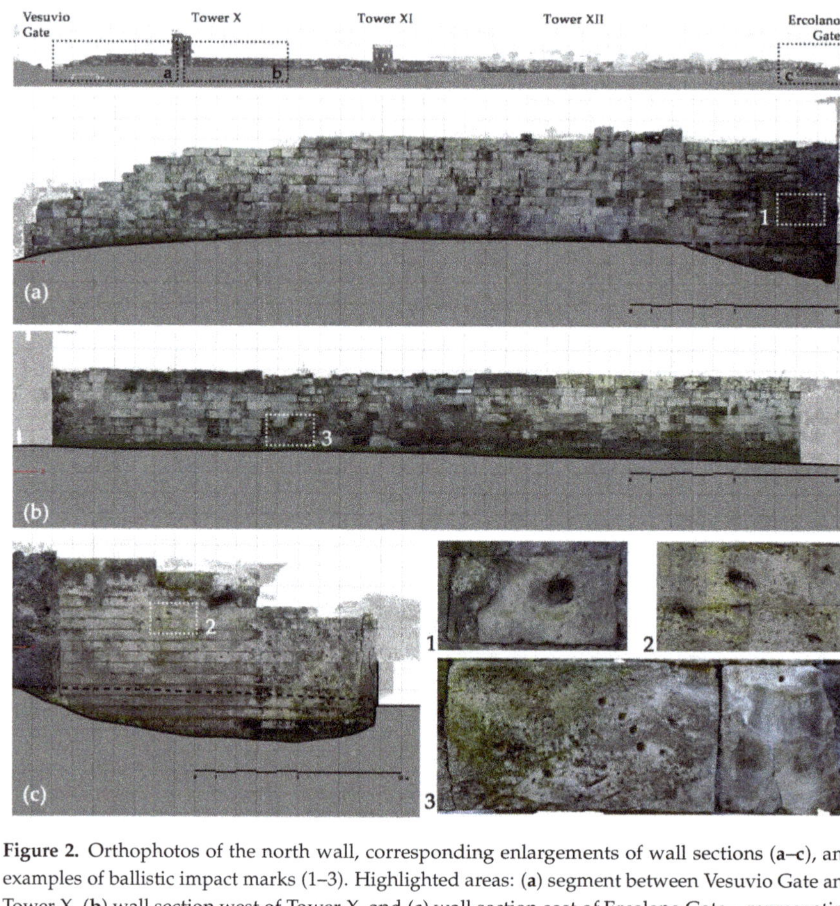

Figure 2. Orthophotos of the north wall, corresponding enlargements of wall sections (**a**–**c**), and examples of ballistic impact marks (1–3). Highlighted areas: (**a**) segment between Vesuvio Gate and Tower X, (**b**) wall section west of Tower X, and (**c**) wall section east of Ercolano Gate—representing the zones most affected by ballistic impacts: (1) stone balls, (2) lead sling bullets, and (3) metal darts.

The study of possible dart impressions therefore presented a challenge to the authors—one that gradually took shape over the course of on-site campaigns. Direct observation of the walls helped to support working hypotheses based on methodologically sound reasoning and provided new elements for discussion—always bearing in mind the inherent uncertainty that characterizes impact marks made by small, pointed weapons such as the metal-tipped darts used by Republican Roman soldiers.

2. Materials and Methods

The investigation focused on examining possible ballistic traces of the Sullan siege still visible along the northern stretch of Pompeii's city walls, with a strong emphasis on methodological accuracy. Numerous high-resolution 3D datasets were acquired using both image-based and range-based survey techniques. The post-processing phase, typically more time-consuming than on-site acquisition procedures, allowed researchers to overcome the limitations of traditional representation methods such as 2D drawings from direct survey or standard photographic documentation. The point clouds, subsequently meshed and enriched with photometric data, were used to identify, isolate, and extract the most relevant case studies. Specialized software facilitated the transition from visualization to the management of polygonal meshes, enabling the creation of models customized to the specific objectives of the research.

Digital surveying also allowed for the accurate measurement of projectile impact angles on stone surfaces, thanks to terrestrial laser scanning as a general reference for locating case studies. This offers an alternative approach to the so-called graphic error, or approximation, resulting from representations on visual outputs or paper, typically introduced by manual interpretation or by printing devices related to scale [7].

However, not all inaccuracies stem from the instruments themselves [8], and operators may introduce both random and systematic errors [9]. Beyond any holistic theories about the nature of percentage error, in this particular case the correspondence between graphical signs and the physical evidence being examined directly shapes data interpretation by what is represented—and how. Precision is not only a mathematical concept; it also reflects the ability to clearly define the specific portion of reality being analyzed [10] (p. 25).

Already in the 1920s, Van Buren had recorded the presence of small traces—as small as 2 cm—alongside larger, more obvious impacts [2] (p. 110). Nearly a century later, the challenges in interpreting such marks persist, complicated not only by the porous structure of the city wall—which contains both natural and anthropic cavities—but also by fundamental concepts of poliorcetics and the intended function of the weapons involved. Dart-launching devices were, in fact, anti-personnel tools, not intended to breach thick walls but rather to wound enemies. As such, it is reasonable to assume that many of the original traces were located either high up (aimed at defenders on the wall walk behind the battlements) [11] (p. 31) or low down (targeting soldiers exiting through sally ports at the base of towers used during sorties). The former have been lost along with the merlons and the smashed wooden defensive structures mounted for defense purposes that once shielded defenders [1,12], while the latter were buried beneath volcanic deposits which, in some sectors, raised the ground level by over two meters (Figure 3).

That being considered, the field inspection therefore prioritized the identification of quadrangular-shaped indentations that, to our knowledge, have never been documented before—shapes more easily distinguishable from natural porosity as well as from impact marks caused by spherical projectiles, sling bullets, or conical arrowheads. Among the various dart tip types, in fact, the most commonly used were hot-forged pyramidal iron heads, theoretically capable of inflicting severe damage comparable to that of spherical missiles, but with reduced deflection upon impact. Examples of iron (and more rarely bronze) arrowheads, mounted on wooden shafts with different tangs and conical or flanged sockets, are preserved in several European museums, including the British Museum [13,14] (Figure 4).

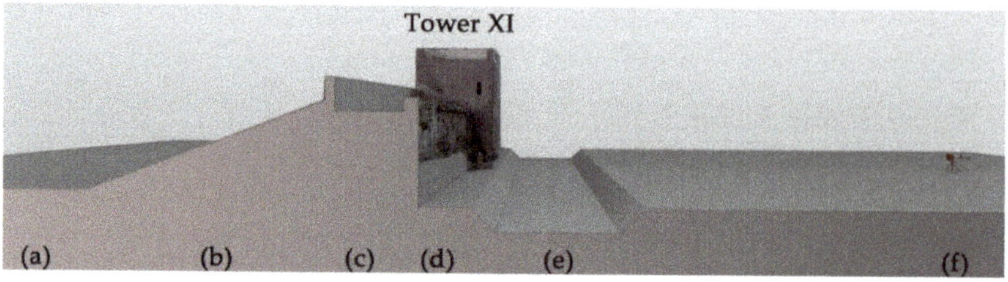

Figure 3. Hypothetical ancient spatial configuration: (**a**) The ancient city of Pompeii; (**b**) The agger (embankment); (**c**) City walls with a wall-walk accessible from the towers; (**d**) Tower sally port; (**e**) The moat; (**f**) The besieging forces and their siege weapons.

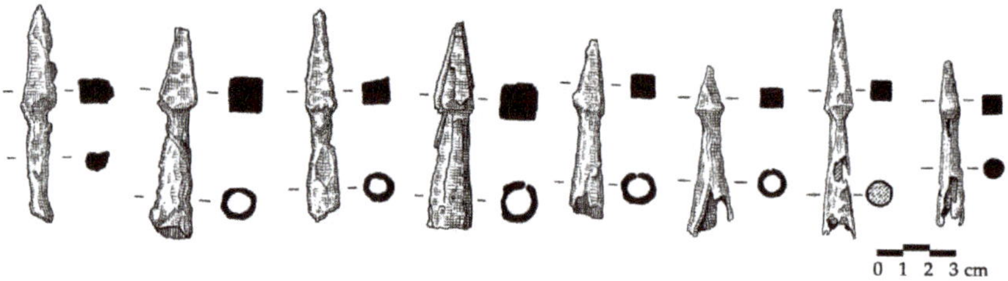

Figure 4. Drawings of iron ballista bolts with pyramidal heads and different sockets, mid-1st century AD, Dorset (UK). British Museum online collection, respective registration nos. 1892,0901.1134-1136-1146-1123-1164-1181; 1960,0405.238-239. Copyright: The Trustees of the British Museum, https://www.britishmuseum.org/collection (accessed on 8 February 2025).

2.1. Impact Survey

Following the above criteria, the first visual inspection focused on square-shaped cavities, some of which were found in clusters. Among these, special attention was given to a fan-shaped, centripetal arrangement consisting of four square depressions and a fifth mark of similar size and geometry. Each cavity measured approximately 24.5 mm in width and reached a maximum depth of about 20 mm (Figure 5). The group is located approximately 89 cm above the current ground level, between Towers X and XI (Figure 6). Confident that these were not natural cavities, it was considered plausible that the impacts at this height were aimed at a moving target—perhaps soldiers moving along the defense line. A second and third similar configuration, although found at different heights, made this fan-shaped group especially worthy of detailed investigation.

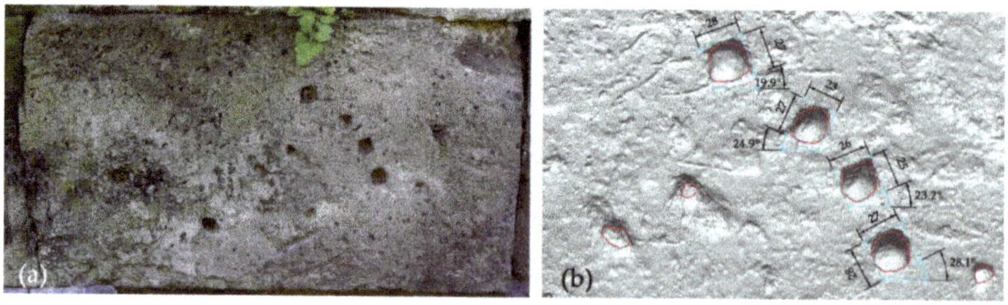

Figure 5. (**a**) Photo of the fan-shaped group in the northern wall, ca. 12 m west of Tower X; (**b**) Close-up on the four square-shaped holes, with dimensions in mm and angles in arc degrees.

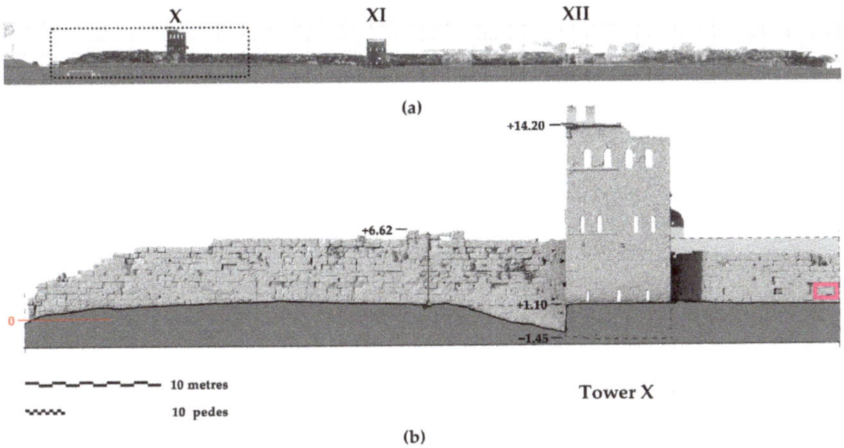

Figure 6. (**a**) Digital model of the surveyed wall section between the Vesuvio and Ercolano Gates. (**b**) Close-up on the selected area of interest, highlighting the case study shown in Figures 5–11 within the pink rectangle (bottom right).

Due to restrictions that prevented placing targets directly on the city walls (see access permission), only non-contact acquisition techniques were considered adequate.

In the first instance, the digital survey was carried out using passive sensors and processed through Structure from Motion (SfM) and Image Matching (IM) techniques. The resolution was chosen according to the specific characteristics of the case, in order to create a high-quality model. The scaling of the model was ensured by using a custom-made scale bar placed within the scene, and a color balance target was also included in the setup to support chromatic balance during post-processing activities (Figure 7).

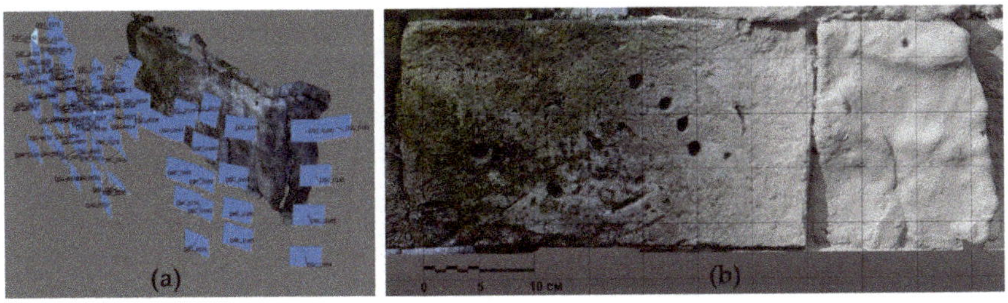

Figure 7. (**a**) SfM model of stone ashlars processed in Agisoft Metashape Pro; (**b**) An orthoimage of the polygonal mesh with blended-map visualizations highlighting the cavities using the Diffuse Map (DM) texture (left), and no texture (right).

Given the favorable lighting conditions at the time of acquisition, no significant issues were encountered; all typical limitations—such as image blur, insufficient depth of field in recessed areas, or poor illumination—were carefully mitigated. In particular, artificial lighting was employed to ensure adequate exposure inside the depressions when necessary.

However, despite all precautions, photogrammetric acquisition faces certain limitations in specific contexts: (i) when the target area is located beyond the operator's physical reach, resulting in incomplete surface coverage and potentially inaccurate reconstructed geometries in those zones; (ii) in the presence of invasive vegetation growing inside cavities, which obstructs a clear view of the occluded surfaces; and iii) in the case of deep holes where light occlusion cannot be fully resolved, even with artificial illumination.

Additionally, since passive techniques create non-metric models, the accuracy of the scaling operation becomes particularly critical, especially when dealing with extremely small or deeply recessed impact marks such as those potentially caused by arrowheads; even small scaling inaccuracies can significantly affect the geometric interpretation of these features and introduce variations in final output [15].

To address these issues, documentation can be supplemented with triangulation-based or structured-light scanners, capable of producing highly dense models with sub-millimetric resolution—within the resolution and depth-of-field limits imposed by the device and the specific logistics of the object.

The device used was a NextEngine Desktop 3D triangulation scanner, which uses MultiStripe Laser Triangulation (MLT): multiple laser stripes are projected onto the surface while a CCD camera (twin 5.0 Megapixel CMOS image sensors) captures the point positions (Figure 8).

The first case study acquired was located at a height that was easily accessible and within the scanner's optimal acquisition range (52–54 cm from the wall surface). During capture, several tests were conducted in the field under different lighting conditions, as ambient light diffusion can interfere with the acquisition phase (these devices are typically designed for indoor use). To fully document the four marks with sufficient overlap for proper alignment (each scan covering roughly an A4-sized area), 15 range maps were acquired (each with a scan time of approximately five minutes depending on the settings).

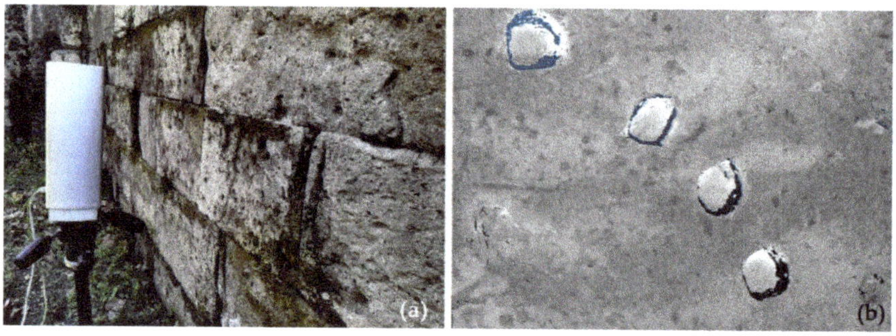

Figure 8. (**a**) Acquisition of a range map with the NextEngine scanner; (**b**) Close-up of a single range map.

2.2. Creation of 3D Models

The scan alignment process was performed using Geomagic Design X software. Once all scans were imported (in mm), they were registered using the N-points method, selecting homologous points on recognizable features across two different overlapping scans. This was followed by a global fine registration process, repeated until all available range maps were properly aligned (Figure 9).

After the initial alignment, cross-sections were extracted every 10 mm to ensure that the deviation between the models remained within acceptable thresholds. This helped validate the central area containing the four features of interest and allowed for a direct deviation comparison between the two meshes: one from SfM and the other from the NextEngine scanner.

Figure 9. (**a**) Step of the N-points alignment process between two scans; (**b**) Final result before merging of the scans: different colors represent different scans.

As a result of the survey campaign, for this representative case study, three distinct digital models were available (Figure 10); all models were located using the same local coordinate system established by the TLS reference model, although typically, each individual model copy has its pivot point at the barycenter to facilitate visualization:

- A general polygonal mesh derived from a portion of the terrestrial laser scanner point cloud (exported in *.ptx format). After standard editing and topological error correction, the mesh had an average edge length of 3.49 mm;
- A photogrammetric polygonal model with an average mesh edge length of 0.83 mm;
- A high-resolution model obtained by registering and merging the 15 range maps acquired with the NextEngine scanner, with an average mesh edge length of 0.41 mm.

The varying levels of detail across the models reflect differences in acquisition technique, device resolution, and optimization. Each model served a distinct purpose: the general TLS mesh provided a spatial reference framework for positioning the case studies along the city walls; the SfM mesh offered detailed surface documentation, including texture data; and the MLT scanner mesh supported high-precision analysis of very small and deeply re-entrant indentations.

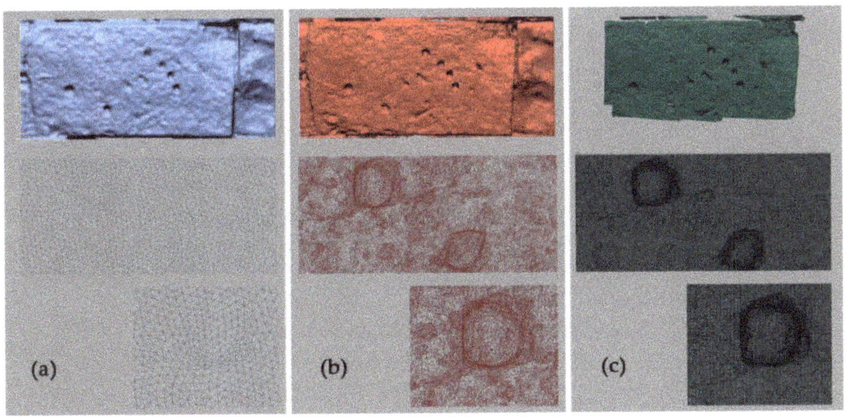

Figure 10. Three-dimensional digital models of the case study area with polygonal mesh details: (**a**) TLS survey; (**b**) SfM survey; (**c**) Next Engine scanner survey.

The resulting polygonal mesh was then carefully analyzed to assess both morphology and dimensional features. Multiple cross-sections were extracted using three reference planes in order to evaluate shape similarity and determine whether the marks might have been caused by the same type of projectile (Figure 11).

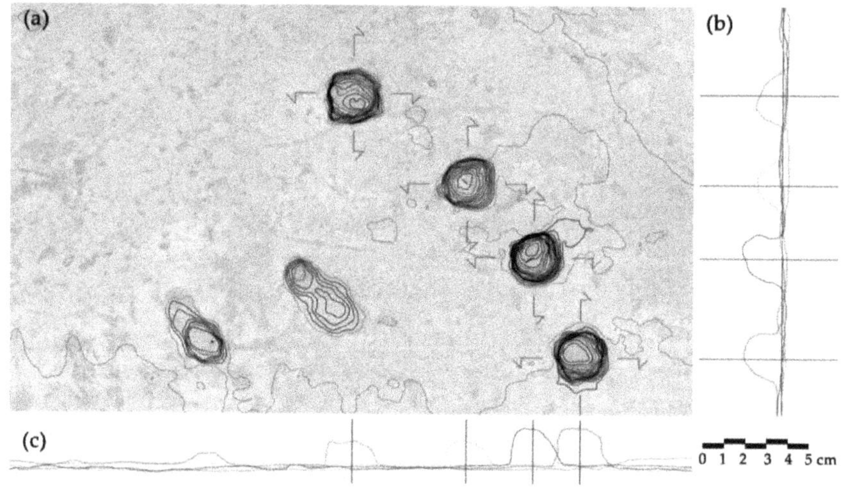

Figure 11. (a) Elevation view, (b) vertical cross-sections, and (c) horizontal cross-sections of the mesh model with contour lines to enhance perception of hollow areas.

3. Results: Virtual Casts and Reverse Modeling

One of the most immediate applications of a polygonal model is the ability to generate virtual casts that replicate the scanned surface. By inverting the surface normals, it is possible to create a negative mold, from which a positive highly detailed digital replica of the original object can be derived [16].

There are, however, certain aspects to be taken into account regarding the degree of approximation inherent to such models, and that the survey captures the object only as it exists at the specific moment of documentation, i.e., the current state of preservation of the surface at a given point in time. Over time, significant alterations have definitely affected the morphology of the impacted surfaces—from the initial damage caused by the Sullan artillery strike, to the volcanic eruption of Vesuvio nearly a century later (79 AD), and the subsequent excavations of the 20th century. On the one hand, the current morphology of the hole is the result not only of the impact, but also of subsequent erosion and micro-fragmentation due to weathering; on the other hand, the presence of patinas and surface encrustations may slightly alter the reliability of the cavity's configuration and measurements. These factors introduce a degree of uncertainty that must be considered in the interpretation of the data, although to a limited extent relative to our purposes. Yet, these surviving marks are invaluable vestiges, offering a rare glimpse into the destructive power of ancient elastic torsion artillery systems.

Dealing with the possible reconstruction of blunt objects through reverse modeling operations starting from effects on the walls, a fundamental step consists in analyzing the available mesh of the documented cavities for reconstruction purposes, as already done in the case of spheroidal cavity marks. Moreover, original shapes of the projectiles can be inferred from surviving archaeological specimens, such as pyramidal metallic arrowheads preserved in museum collections and available through online archives. These can serve as references, at least in terms of their known measurements and proportions, to guide the reconstruction of the object's solid geometry.

In this case, direct reconstruction of their original geometry using fully automated best-fitting tools applied to the cavities was not feasible. Therefore, different approaches were tested to reconstruct a representative dart shape, either through semi-automatic tools or by manual modeling carried out by the operator to define the most plausible morphological hypothesis.

The applied workflows were as follows:

1. Surface-Based Reverse Modeling (SBRM)—an automated method for finding the shape of the cavity and the most likely axis of penetration;
2. Direct Geometric Modeling (DGM)—a method for geometrically reconstructing the dart from cross-sections based on known dimensions and proportions.

3.1. *Surface-Based Reverse Modeling (SBRM)*

The high-resolution mesh derived from the MLT scanner served as the basis for extracting significant geometric information from the cavities (Figure 12). Although the model was already metric and detailed, it still contained minor surface irregularities—such as small protrusions, localized noise, or debris—which could affect the interpretation of morphological data. For this reason, the mesh was subjected to a light defeaturing and smart smoothing process, aimed at reducing noise and irregularities, and improving the uniformity of selected polygon faces (within a specified deviation range), while preserving morphologically relevant features.

This pre-processing step is crucial before using automatic surface region detection tools, as smoother and more uniform meshes yield less fragmented segmentation results. The procedure segments the mesh based on curvature variation and geometrical continuity, assigning distinct color codes to each detected region.

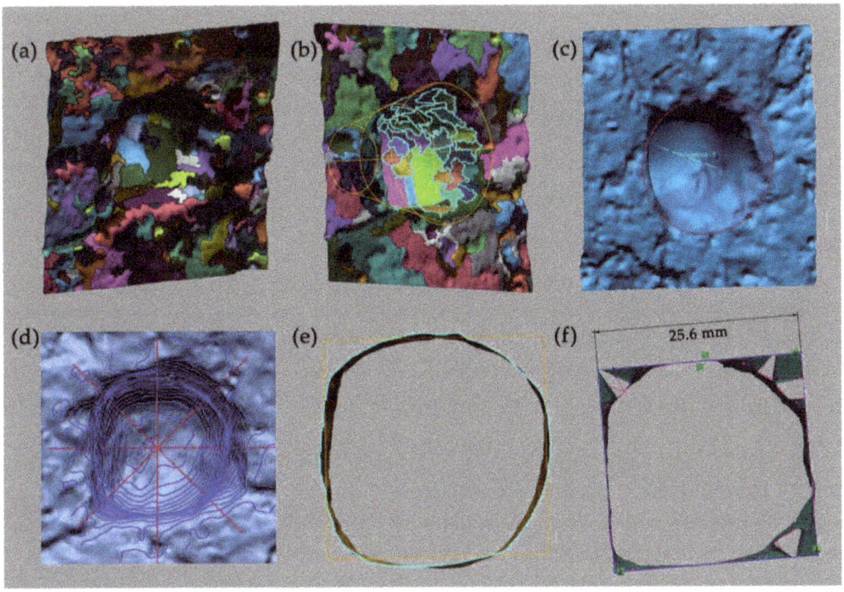

Figure 12. (**a**) Surface region detection; (**b**) Identification of best-fitting geometric primitives by selecting specific regions; (**c**) Axis of penetration; (**d**) Radial or orthogonal sections of the cavity; (**e**) Selection of sections; (**f**) Rectification into best-fitting square shapes.

Operators can adjust sensitivity or manually re-segment key areas using cutting polylines, in order to increase or decrease the number of identified regions. This process facilitated the identification of best-fitting geometric primitives by selecting only the regions that corresponded to the impact area.

Since the outcome depends on the region selection, it is necessary to define in advance which geometric shape is to be extracted and carefully choose which regions to include in the computation. Moreover, only a limited set of simple geometric primitives is supported by the automatic extraction tool. Therefore, it may be more effective to first examine a similar shape—for instance a cone that approximates the pointed pyramidal shape of the arrowhead and conceived as enveloping the solid to be reconstructed—and then define the most probable impact axis (possibly coinciding with the axis of the automatically extrapolated cone), which is not always visible to the naked eye, especially in shallow cavities.

Once identified, the geometric regions were analyzed to determine which primitive shapes (e.g., cones) best fit the selected surfaces. In this way, a first hypothetical 3D shape could be obtained, corresponding to the best-fitting geometry of the impacting object.

The analysis also allowed for the identification of the most likely axis of penetration, often coinciding with the axis of symmetry of the best-fitting primitive. From this axis, it was then possible to extract multiple radial or orthogonal sections from the cavity and use them to better define the internal profile of the impact mark.

To improve legibility and support visual comparisons, the extracted sections can be cleaned of irregular outlines and, in some cases, manually adjusted to follow an idealized quadrangular or conical path.

With such sections, either by 3D sketches on the mesh or by switching to automatic contour editing, which aims to rectify them into a square shape, it is possible to obtain planar geometries from which surfaces or composite solids can be created to reconstruct the object.

The contour editing tool allows the adjustment of mesh boundaries to a predefined shape—for example, reshaping the contour of polygons to have equal sides (in this case, four sides to obtain the best-fitting square section) and modifying the mesh accordingly. At this point, the best fitting section can be determined through sketching.

With both methods, the side length of the square base of the reconstructed truncated pyramid in the stone ashlar was found to be approximately 2.5 cm, equal to one Roman *uncia* (1/12 of *pes*, i.e., 2.465 cm, given that 1 *pes* = 29.581 cm; cf. [17] (p. 89)).

3.2. Direct Geometric Modeling (DGM)

Although more time-consuming, manual modeling offers valuable control over the reconstruction process, especially when working with complex shapes or incomplete cavities where fully automated solutions are not always sufficient.

Therefore, in addition to the semi-automatic workflow, a more interpretative approach was adopted to reconstruct the dart geometry based on dimensional observations and available references (Figure 13).

Using CAD tools, the operator can draw a series of 2D sketches on various planes (lines, n-sided polygons, etc.), employing basic functions such as trimming, extending, or mirroring, and based on measurements taken from the existing specimens (e.g., arrowheads from British Museum collections), particularly those with quadrangular cross-sections. These sketches are then used to generate solid shapes through lofting, extrusion, or revolution operations of closed profiles, producing simplified yet plausible representations of the projectile.

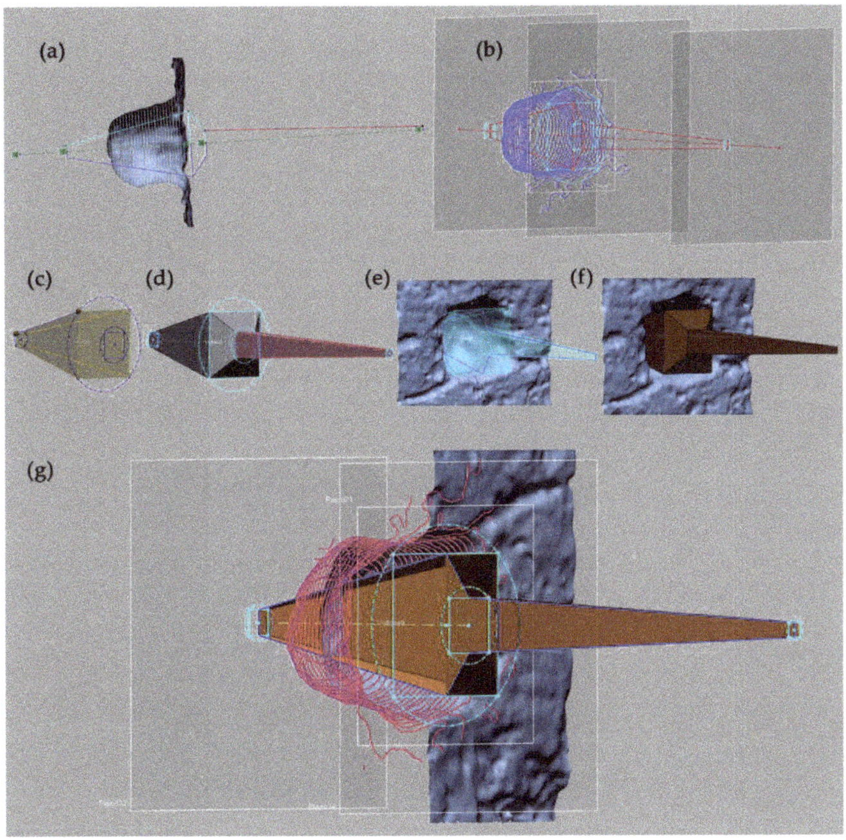

Figure 13. (**a**,**b**) Two-dimensional sketches of the arrowhead drawn on reference planes; (**c**–**f**) Creation of the solid geometry by extruding or lofting base sketches; (**g**) Comparison of the modeled object with reference profiles.

Where necessary, slight adjustments were made to refine the model's fit—such as offsetting surfaces, scaling components, or reshaping sections—to better align the geometry with measured data and hypothesized impact dynamics.

This method allowed for greater flexibility, enabling real-time modifications of the model based on morphological analogies with museum specimens, and facilitating an iterative validation process comparing the reconstructed object with the recorded cavity geometry.

4. Discussion

Confirming, overall, the potential of rapidly advancing digital technologies, already in use for some time [18,19], the integration of these surveying techniques has allowed for the generation of accurate and detailed virtual reconstructions of ballistic impact marks, even in the case of small, geometrically complex craters. The integrated pipeline of advanced point cloud processing, mesh generation, and reverse modeling enables the creation of both positive and negative virtual casts, documenting the preserved morphology of the wall surface at the moment of the survey.

TLS, SfM, and IM not only support the reverse modeling stages but also facilitate the accurate registration of shape and size, depth of penetration, and incidence angle of projectile impacts (e.g., original launch trajectory or angle of strike) in relation to a horizontal reference. The spatial consistency of the dataset makes it possible to evaluate each feature in its spatial context, with georeferencing of these impacts along the northern curtain wall to reconstruct their original height and spatial distribution within the ancient defensive context, contributing to broader polemological interpretations and helping to assess tactical, strategic, or interdisciplinary aspects in the study of ancient siege warfare.

High-detail 3D reconstructions, derived from meshes with varying edge resolutions, can subsequently be converted into numerically reliable and optionally 3D-printable solid models. These reconstructions now serve as a scientifically robust and reliable record. A structured classification protocol will allow these data to be integrated into a shared, interoperable workflow for future comparative studies.

Furthermore, reverse engineering processes will allow hypotheses about projectile shape and behavior to be tested in future simulations. In this way, 3D modeling becomes both a descriptive and predictive tool, capable of informing future reconstructions of the dart-launching catapults believed to have been used during the Sullan siege.

Moreover, the geometric reconstruction of possible weapons and projectiles will benefit from advances in digital modeling, which in recent years have also been applied in the context of lost or undocumented artifacts, especially when only a few historical images survive. Advances in deep learning and artificial intelligence now allow 3D reconstruction from minimal inputs [20,21], supporting new forms of digital restitution even in the absence of complete material data (Figure 14). These methods, while experimental, demonstrate the growing potential of combining archaeological data with AI-based modeling.

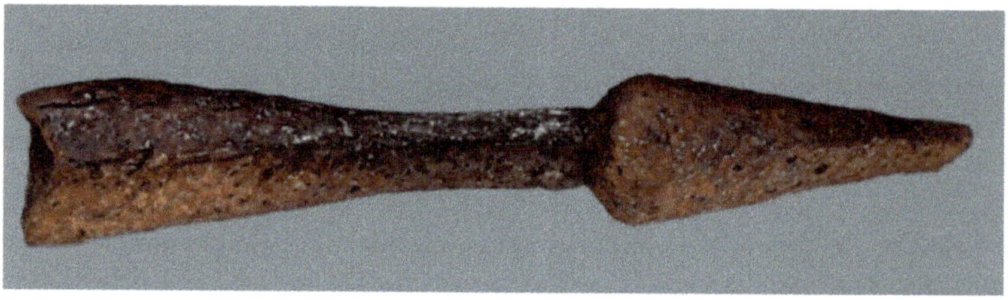

Figure 14. AI-generated 3D model of an iron ballista bolt with a small, neatly made pyramidal head and a conical socket with a rivet near the mouth (mid 1st century AD, Dorset, UK), created from an image in the British Museum online collection (registration no. 1892,0901.1144), using the free version of Vizcom.ai (https://app.vizcom.ai/, accessed on 21 October 2024).

5. Conclusions

At the end of this initial phase of digital documentation and critical analysis, carried out during this first year, the authors propose a hypothesis that—although requiring further verification—may offer an exciting new perspective: the possible identification of impact traces left by a repeating catapult, known in ancient sources as the *polybolos*. This weapon, attributed to the ingenuity of Dionysius of Alexandria (an engineer active in the arsenal of Rhodes), is described in detail as the first known automatic missile launcher in history (Philo of Byzantium, *Belopoeica*, 73.34).

The "multi-launcher" differed from traditional elastic torsion catapults in one key aspect: it could launch multiple darts in rapid succession without the need for manual reloading. It represented the peak of ancient mechanical artillery—a sophisticated, semi-automatic system that may have been known to the Roman military elite during the period.

The fan-shaped arrangement of the four square-shaped impact marks—similar in size and depth and regularly spaced—could suggest a repetitive firing mechanism that the authors identified. Their slight angular rotation relative to a common reference axis, along with the presence of a fifth, slightly misaligned imprint, strengthens the hypothesis that the impacts could have been generated by a multi-shot system [22] (Figure 15), which could launch in rapid succession, i.e., the *polybolos*, meticulously described by Philo of Byzantium (3rd century BC).

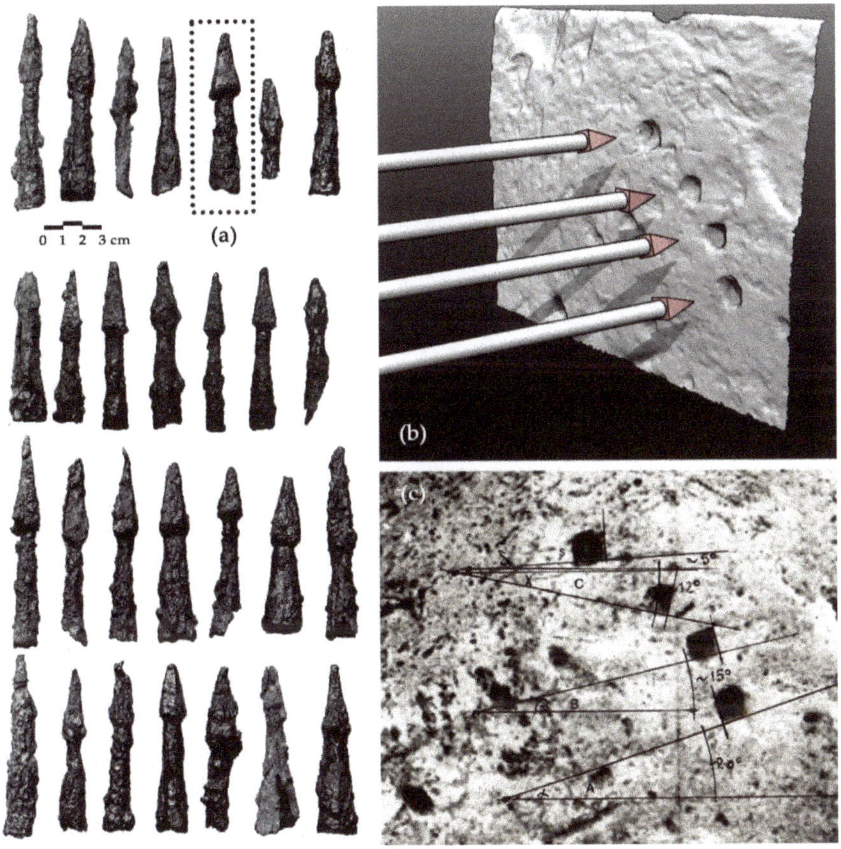

Figure 15. (**a**) Photo of various iron ballista bolts with pyramidal heads and various sockets, mid-1st century AD, Dorset (UK). British Museum online collection, registration nos. 1892,0901.1138/1129. Copyright: The Trustees of the British Museum. The metal tip highlighted in the dotted rectangle was selected for 3D reconstruction. (**b**) Hypothetical reconstruction of multiple firings of metal-tipped darts onto the digitally surveyed surface. (**c**) Photograph of the fan-shaped group, with approximate identification of the inclinations that directed the digital survey.

The absence of a technical glossary makes the original treatise—in ancient Greek—unintelligible to non-specialists. Expert engineers fluent in classical language helped interpret passages deliberately vague at key points. Their critical interpretation enabled a reconstruction compatible with the materials and technologies of the time for the multi-launcher depicted in the 1:20-scale plates accompanying the German translation by Diels and Schramm [23] (chap. 51, 58, "Mehrlader nach Philon").

The literal descriptions in some sections of the Greek text suggest that Philo—supposed disciple of Ctesibius [24]—handled or witnessed the *polybolos* in operation while in besieged Rhodes (305–304 BC). Later scholars, including Russo, reinterpreted the key passages to deepen our understanding of its mechanisms [25], and accounts celebrate its civilian uses in the Imperial period.

The comparison with nineteenth-century weapons such as the Gatling gun, patented in 1862, which used multiple rotating barrels to produce similar firing patterns, invites reflection on the level of mechanical engineering already achieved in antiquity.

Although uncertainties still envelop these preliminary hypotheses, the singular fan-shaped grouping of roughly square indentations defies any logic of human or natural damage. One isolated case might be fortuitous, two similar configurations coincidental—but three analogous instances compel serious consideration in hoplological science.

It is highly unlikely that multiple archers could have produced such uniformly spaced marks independently, nor does it seem plausible that different torsion catapults could have replicated such identical tension and aim under battlefield conditions. The hypothesis of a weapon capable of obtaining similar results, assuming an identical tension of the bowstring and therefore identical loading of the elastic strands, remains difficult if not nearly impossible to reconcile with the properties of conventional elastic-torsion mechanisms. Hence, the idea of a single, stable device producing such a regular dispersion is more consistent with the known characteristics of a multi-launcher: the authors reasonably assume they have detected proof of the *polybolos* in action, which could possibly explain the slight angular rotation of these indentations, similar for shape and dimensions, in relation to a horizontal reference. Significantly, the square marks on the stone appear at regular spacing inwards—caused by darts falling into the hopper with each release of the firing mechanism.

Despite alterations to the stone surface—including weathering, restoration, archaeological excavation, and even modern vandalism—the surviving ballistic traces retain their diagnostic potential. Their preservation, in spite of volcanic destruction, Roman repairs, and environmental degradation, marks them as rare physical evidence of the power of ancient projectile weaponry.

According to ancient sources, Sulla—who governed the province of Asia during the First Mithridatic War—would likely have had access to advanced Hellenistic artillery [26]. It is therefore not unreasonable to suggest that he may have deployed a *polybolos* during the siege of Pompeii. While this remains a working hypothesis, it opens new avenues for interdisciplinary investigation currently in progress [27].

Author Contributions: Conceptualization, A.R.; methodology, all authors; software, S.B.; formal analysis and investigation, A.R.; writing—original draft preparation, A.R.; writing—review and editing, S.B.; supervision, A.R.; funding acquisition, A.R.; photogrammetric survey: A.R., S.B. and other members of the UniVanvitelli Team; laser scanner survey, meshing and 3D model elaboration: S.B. Material preparation was performed with different software, in particular Agisoft Metashape elaboration: S.B.; Leica Geosystems Cyclone TLS cloud processing and registration: S.B.; 3D Systems Geomagic Design X meshing and processing of polygonal mesh: S.B.; Adobe Photoshop CC elaboration of figures: S.B. All authors have read and agreed to the published version of the manuscript.

Funding: This research was supported by the project "SCORPiò-NIDI", CUP B53D23022100006 (DD n. 1012/2023), funded by the Italian Ministry of Research under the PRIN (call DD n. 104/2022) funding initiative.

Institutional Review Board Statement: Not applicable.

Informed Consent Statement: Not applicable.

Data Availability Statement: Research data are available upon request.

Acknowledgments: The authors would like to express their gratitude to the management and the appointed officials of the offices of the Pompeii Archaeological Park for granting authorizations for site access and survey operations.

Conflicts of Interest: The authors declare no conflicts of interest.

Abbreviations

The following abbreviations are used in this manuscript:

3D	Three-Dimensional
AI	Artificial Intelligence
DM	Diffuse Map
IM	Image Matching
DGM	Direct Geometric Modeling
MLT	MultiStripe Laser Triangulation
OM	Occlusion Map
SBRM	Surface-Based Reverse Modeling
SfM	Structure from Motion
TLS	Terrestrial Laser Scanner

References

1. Maiuri, A. *Introduzione allo Studio di Pompei, a Cura di G. Oscar Onorato, Corso di Antichità Pompeiane ed Ercolanesi 1942–43*; GUF: Napoli, Italy, 1943.
2. Van Buren, A.W. Further Studies in Pompeian Archaeology. *Mem. Am. Acad. Rome* **1925**, *5*, 103–113. [CrossRef]
3. Van Buren, A.W. Further Pompeian Studies. *Mem. Am. Acad. Rome* **1932**, *10*, 7–54. [CrossRef]
4. Burns, M. Pompeii under Siege: A Missile Assemblage from the Social War. *J. Rom. Mil. Equip.* **2015**, *14*, 1–10.
5. Langella, A.; Bish, D.L.; Calcaterra, D.; Cappelletti, P.; Cerri, G.; Colella, A.; Graziano, S.F.; Papa, L.; Perrotta, A.; Scarpati, C.; et al. L'Ignimbrite Campana (IC). In *Le Pietre Storiche della Campania dall'oblio alla Riscoperta*; de Gennaro, M., Calcaterra, D., Langella, A., Eds.; Luciano Editore: Napoli, Italy, 2013; pp. 155–177. ISBN 88-6026-182-3.
6. Kastenmeier, P.; Di Maio, G.; Balassone, G.; Boni, M.; Joachimski, M.; Mondillo, N. The Source of Stone Building Materials from the Pompeii Archaeological Area and Its Surroundings. *Period. Di Mineralogia* **2010**, *87*, 38–58. [CrossRef]
7. *ISO/IEC Guide 98-3*; 2008 Uncertainty of Measurement—Part 3: Guide to the Expression of Uncertainty in Measurement (GUM:1995). International Organization for Standardization: Geneva, Switzerland, 2008.
8. Taylor, J.R. *Introduzione all'Analisi Degli Errori. Lo Studio Delle Incertezze Nelle Misure Fisiche*, 3rd ed.; Zanichelli: Bologna, Italy, 2023; ISBN 978-88-08-39966-3.
9. Heisenberg, W. *Fisica e Oltre. Incontri Con i Protagonisti 1920–1965*, 6th ed.; Bollati Boringhieri: Torino, Italy, 2013; ISBN 978-88-339-2416-8.
10. De Rubertis, R. *Il Disegno dell'Architettura*; NIS: Rome, Italy, 1994; ISBN 978-88-430-0272-6.
11. Russo, F.; Russo, F. *89 a.C. Assedio a Pompei: La Dinamica e Le Tecnologie Belliche della Conquista Sillana di Pompei*; edizioni Flavius: Pompei, Italy, 2005; ISBN 88-88419-32-2.
12. Vitucci, G. (Ed.) *La Guerra Giudaica di Giuseppe Flavio*; Mondadori: Milano, Italy, 1989; ISBN 88-04-32627-1.
13. Payne, G. (Ed.) *Catalogue of the Museum of Local Antiquities Collected by Mr Henry Durden of Blandford, Dorsetshire*; South Counties Press: Lewes, DE, USA, 1892.
14. Manning, W.H. *Catalogue of the Romano-British Iron Tools, Fittings and Weapons in the British Museum*; Published for the Trustees of the British Museum by British Museum Publications: London, UK, 1985; ISBN 0-7141-1370-0.
15. Guidi, G.; Micoli, L.L.; Gonizzi Barsanti, S.; Brennan, M.; Frischer, B. Image-Based 3D Capture of Cultural Heritage Artifacts an Experimental Study about 3D Data Quality. In Proceedings of the 2015 Digital Heritage, Granada, Spain, 28 September–2 October 2015; Volume 2, pp. 321–324.
16. Helle, R.H.; Lemu, H.G. A Case Study on Use of 3D Scanning for Reverse Engineering and Quality Control. *Mater. Today Proc.* **2021**, *45*, 5255–5262. [CrossRef]
17. Lazzarini, M. Metrologia romana. In *Conimbriga*; Universidade de Coimbra: Coimbra, Portugal, 1965; Volume IV.
18. Remondino, F.; El-Hakim, S. Image-Based 3D Modelling: A Review. *Photogramm. Rec.* **2006**, *21*, 269–291. [CrossRef]
19. Guidi, G.; Russo, M.; Beraldin, J.-A. *Acquisizione 3D e Modellazione Poligonale*; McGraw-Hill Education: New York, NY, USA, 2010.
20. Kniaz, V.V.; Remondino, F.; Knyaz, V.A. Generative Adversarial Networks for Single Photo 3D Reconstruction. *Int. Arch. Photogramm. Remote Sens. Spat. Inf. Sci.* **2019**, *XLII-2-W9*, 403–408. [CrossRef]
21. Fu, K.; Peng, J.; He, Q.; Zhang, H. Single Image 3D Object Reconstruction Based on Deep Learning: A Review. *Multimed. Tools Appl.* **2021**, *80*, 463–498. [CrossRef]

22. Rossi, A.; Gonizzi Barsanti, S.; Bertacchi, S. Use of *Polybolos* on the City Walls of Ancient Pompeii: Assessment on the Anthropic Cavities. *Nexus Netw. J.* **2024**, *27*, 1–30. [CrossRef]
23. Diels, H.; Schramm, E.A. *Philons Belopoiika, viertes Buch der Mechanik*; Akademie der Wissenschaften in Kommission bei G. Reimer: Berlin, Germany, 1919.
24. Drachmann, A.G. *Ktesibios, Philon and Heron: A Study in Ancient Pneumatics*; Ejnar Munksgaard: Copenaghen, Denmark, 1948.
25. Rossi, C.; Russo, F. *Ancient Engineers' Inventions*; History of Mechanism and Machine Science; Springer International Publishing: Cham, Switzerland, 2017; Volume 33, ISBN 978-3-319-44475-8.
26. Piganiol, A. *Le Conquiste Dei Romani. Fondazione e Ascesa di Una Grande Civiltà*; Il Saggiatore: Milano, Italy, 2010; ISBN 978-88-565-0162-9.
27. Rossi, A.; Formicola, C.; Gonizzi Barsanti, S. Ingegna Romana. Dalle fonti ai modelli, dai reperti alle ricostruzioni. *Diségno* **2024**, *14*, 229–238. [CrossRef]

engineering proceedings

MDPI

Proceeding Paper

(Im)material Casts from the Sullan Period †

Claudio Formicola, Silvia Bertacchi and Adriana Rossi *

Department of Engineering, Università Degli Studi Della Campania Luigi Vanvitelli, 81031 Aversa, Italy; claudio.formicola@unicampania.it (C.F.); silvia.bertacchi@unicampania.it (S.B.)
* Correspondence: adriana.rossi@unicampania.it
† Presented at the Conference "Discovering Pompeii: From Effects to Causes—From Surveying to the Reconstructions of Ballistae and Scorpiones", Aversa, Italy, 27 February 2025.

Abstract: Thanks to Pompeii's burial under Vesuvio's 79 AD eruption deposits, the ballistic imprints on its northern defensive perimeter are uniquely attributable to Sulla's siege of 89 BC. These impact marks were digitally documented using integrated survey techniques and custom pipelines. The virtual casts generated—dimensionally accurate, high-resolution surface replicas—serve as key inputs for the reverse-modeling of damage craters, supporting terminal ballistics analyses. Two case studies—a stone projectile cavity and fan-shaped dart impressions—were 3D-printed at 1:1 scale. Prototype casting thus emerges as a cultural asset and rapidly updatable component of a dynamic data ecosystem, inclusive of users with disabilities.

Keywords: 3D printing; accessibility; ballistic imprints; digital heritage; elastic torsion launcher; life-size printing; Sulla's siege; rapid prototyping; reverse modeling; virtual casts

1. Introduction

Handling 3D-printed digital casts—prototypes of the ballistic imprints discovered along the perimeter of ancient Pompeii [1]—is striking. Although removed from their original context, the casts reveal the surface roughness and dimensions that evoke the force of impact, recalling the dramatic descriptions by Titus Flavius Josephus, who witnessed the Siege of Jerusalem in 70 AD [2].

The artillery employed on that occasion was essentially identical to that used by Lucius Cornelius Sulla when he took Pompeii in 89 BC [3]. The surviving stretch of wall, preserved beneath volcanic deposits from the 79 AD eruption and only excavated in the early twentieth century, now stands as the sole definitive evidence of the lethal capabilities of ballistae and catapults before Christ; the craters in the stone ashlars, caused by the violent impacts of limestone spherical projectiles and metallic arrows (Figure 1), cannot be mistaken for damage sustained during World War II bombings, which have been documented in detail [4].

Thus, confident in their secure dating, the direct examination of these cavities—grounded in foundational knowledge of Republican and Imperial poliorcetics—allows us to identify features that, despite ancient restorations and a century of atmospheric exposure, form the basis for sizing elastic torsion launchers, of which only a few fragmentary remains exist [5,6].

The design principles outlined by Philo of Byzantium (*Belopoeica*, 3rd century BC) and later codified by Vitruvius in Book X of *De Architectura* (1st century AD) enable the calculation of siege engine components according to a standard module (Figure 2a,b)—the side of the "modiolus" (little module) equals the diameter of the flanges that secure the elastic bundles to the frame, a dimension from which ancient authors derived weapon power [7].

Figure 1. Images of the ballistic imprints (circular marks caused by stone projectiles) visible on the northern city walls of Pompeii. Photo 2016 (MIBACT-SSBA-PES PROTO-ARCH 0006225 14 April 2016 CI. 28.13.07/6).

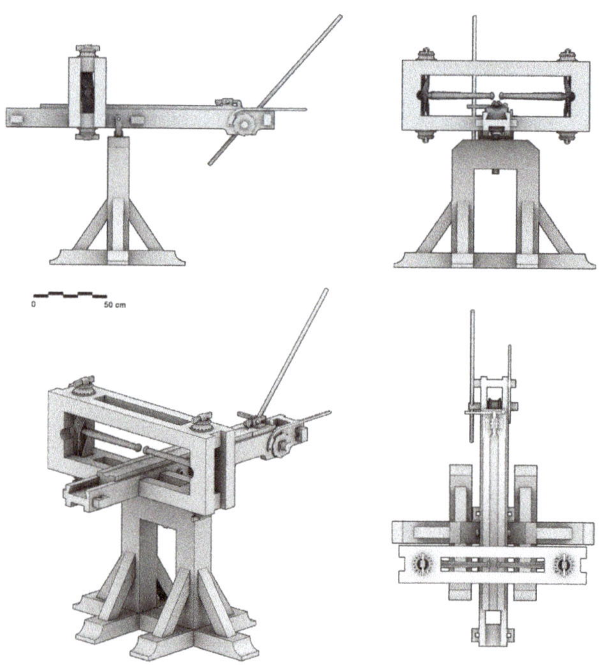

(a)

Figure 2. *Cont.*

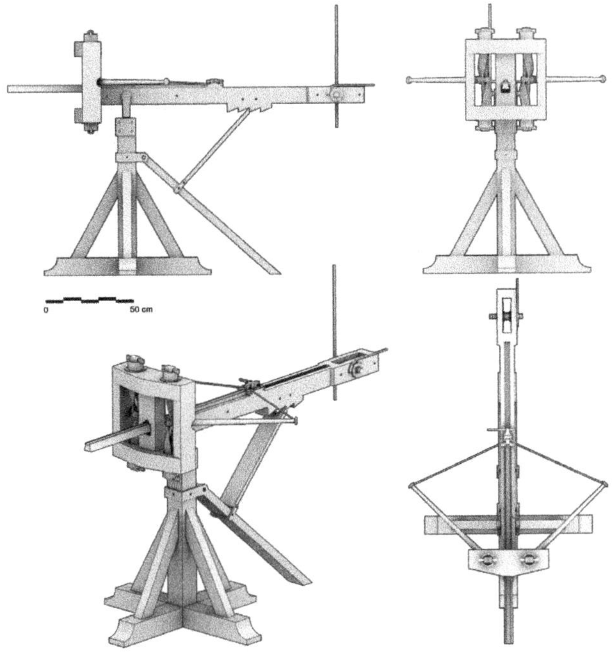

(b)

Figure 2. Orthographic and axonometric views of the 3D model of a Roman weapon: (**a**) ballista; (**b**) scorpion. Models by C.F.

By measuring the bundle diameter, one can obtain firing parameters that can subsequently be partially validated through reverse engineering processes (Figure 3).

The digital process used to capture and reconstruct the virtual casts, which are subsequently prototyped, integrates phases that are traditionally separate. Cyclic and interactive processes of analysis and synthesis follow one another within a single pipeline for data acquisition. These processes yield a collection of information useful for transferring content and fostering collaboration through Information and Communications Technology (ICT). Sharing these ontological structures provides users with informational platforms and (im)material spaces [8] suitable for expanding their perceptual boundaries (Mixed Reality, MR). By introducing digital elements, advanced technologies promote not only visual inclusion but also the cognitive inclusion of users (Extended Reality, XR). It is therefore necessary to reconcile, on the same surveyed and processed models, the precision and accuracy required for documentation and study with the exemplification demanded by the sharing of results in progress and for future reference. Dedicated and interoperable platforms (Common Data Environment, CDE) widely engage users, facilitating access to resources for educational, social, and economic programs.

The primary question this paper seeks to address concerns a critical phase of the unique, cyclical and interactive process—the physical prototyping of digital casts selected as case studies, which represents a significant mode of operation and serves as a an opportunity for cultural exchange between research, education and the so-called Third Mission.

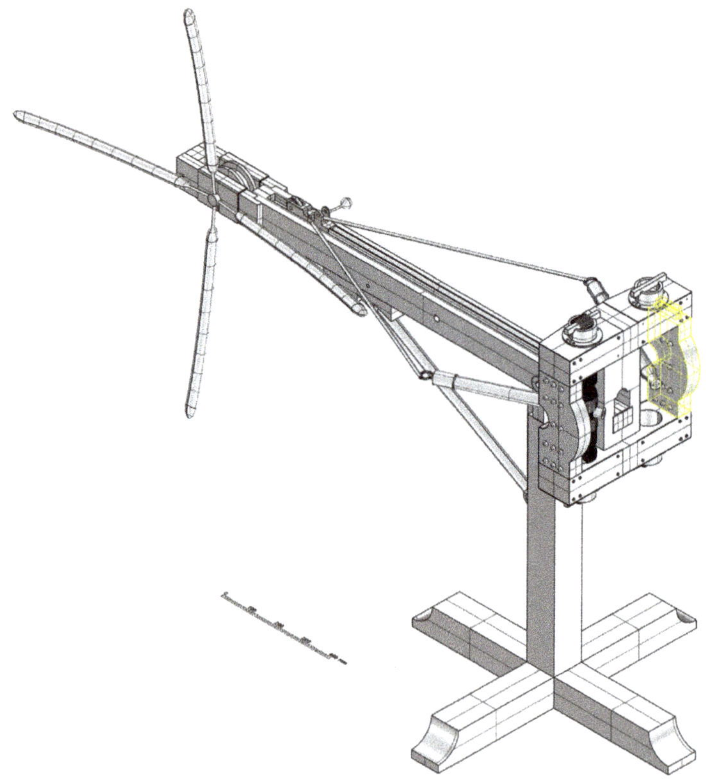

Figure 3. An axonometric view of the 3D model of a Roman weapon, called the *scorpio*. Model and rendering by C.F.

2. Materials and Methods

This research is based on graphic outputs derived from visual representations designed to produce digitally accessible content. Both expressive and conceptual—whether iconic or symbolic, survey- or design-based—the graphic results reflect an experience in which apparent antinomies dissolve into action, illustrating the interplay between existing work and future planning. As Migliari (2004) observes [9], it is appropriate to provide the term "drawing" with wider connotations, to encompass the notion of "model". It is certainly a model, as the prototype casts are derived from reality-based 3D processing. The process is a recording of the physical world as it appears at the moment of capture [10].

The output of post-processing operations on point clouds acquired with active and passive sensors are models consisting of polygonal meshes. Exported at resolutions appropriate to their purpose, and free of holes or topological defects, these 3D models are adaptable to various applications, including effective dissemination via 3D printing.

Photogrammetric reconstructions using Structure from Motion (SfM) precede the 3D prints of the impact case studies selected for the analysis. To preserve contextual integrity, the studied cavities were located in the approximately 300 m wall section between Vesuvio Gate and Ercolano Gate. This reference frame was necessary to locate the different elevations relative to the actual ground level and determine each projectile's angle of incidence on the ashlars and the resulting crater geometry. Such data are essential to the subsequent mechanical tests that intend to support terminal ballistics calculations.

The polygonal mesh documents impact features in high detail and in a complete way. When analyzed with dedicated software for automatic surface characterization, it yields geometric primitives that allow us to hypothesize, with reasonable confidence, the most plausible shapes of the studied impacts (reverse modeling, RM).

The Finite Element Method (FEM) and Finite Element Analysis (FEA) will then validate these hypotheses and determine the launching power of the elastic torsion launchers, starting from the most significant and well-preserved examples, chosen because of the high circularity of the profile, diameter size, and depth.

Digital casts, saved in STL format (Stereo Lithography or Standard Tessellation Language), are used to prototype object geometry without preserving color or texture information.

Digital casts—both the negative molds (inverted geometry) and their positive casts (replica geometry of the cavity)—are extracted generally without color texture in the appropriate format for printing, usually the Stereo Lithography or Standard Tessellation Language (STL), and used to materially replicate the geometry of the object.

The use of an analog, continuous model that extends into physical space "provides not only visual perception but also haptic feedback (from the Greek word for Touch), material scent of which it is composed, and acoustic resonance, sharply contrasting with purely digital processing" [11]. In this light, reliable 3D-printed copies of detailed digital casts confirm the advantages of historical maquettes based on technical drawings, even if they are not handmade with fine materials; beyond illustrating [12] (p. 117) or persuading patrons [13] (p. 40, 207), they can verify the structural behavior and coordination of each element [14] (I, pp. 860–862), that is, helping in studying extreme scales [15] (V, 18, p. 467) or guiding construction [16]. Prototyped models thus open up new opportunities [17].

In fact, 3D printing, while offering a choice, albeit a limited one, uses materials that are not perishable and limit cracking, deforming or shrinking over time, and they can be reprinted for purposes of reducing costs and time. In addition, since they derive from digital models, they enable real-time corrections and rapid, low-cost reprints. These rectifications and optimizations can support user inclusion [18]—with proper attention, they compensate for some physical deficits [19,20] or can extract features of real complexity suitable for the understanding and enjoyment for younger or differently abled children.

We therefore ask how and to what extent these case studies can advance the field. In the context wherein case studies serve not as trivial applications but as an opportunity for dialogue, it is crucial to direct acquisition projects that can capture exceptionally irregular surfaces, such as impact craters, and to derive an inverse model that closely approximates reality. Ensuring dimensional and geometric reliability is critical for the quality of the subsequent steps.

3. Results: Virtual Casts and Reverse Modeling

Two types of case studies were surveyed, and accordingly two examples were selected, as follows:

1. A cavity created by a stone projectile, together with its positive cast and the corresponding theoretical projectile (sphere), of which the maximum diameter was deduced from the cavity via geometric analysis of the surface (Figure 4);
2. A series of fan-shaped cavities, likely produced by metal-tipped darts, together with its positive cast and a reconstructed theoretical dart (frustum of a square pyramid), modeled on the proportions of a Roman-period example that is still preserved (Figure 5). Figure 6 illustrates the study of the arrow tip chosen for prototyping.

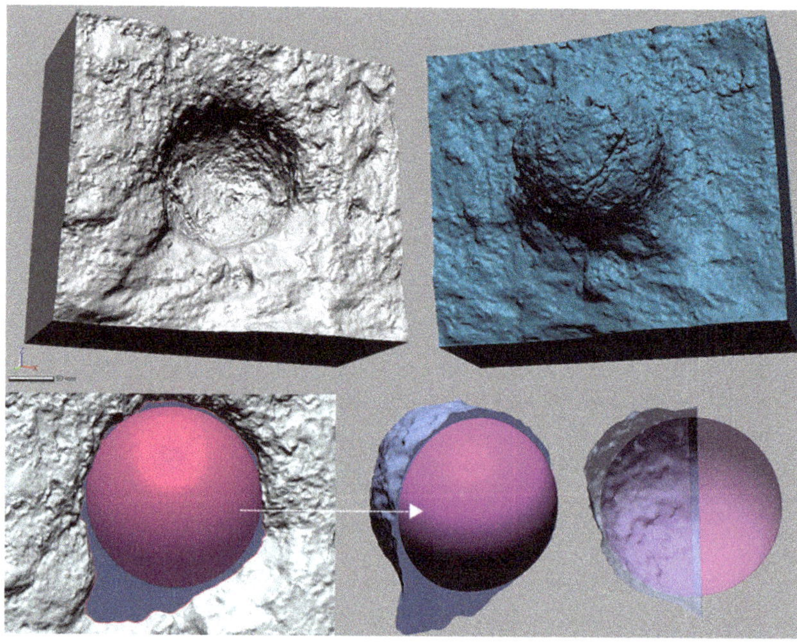

Figure 4. 3D models of the positive cast (**above left**) and the negative mold (**above right**) obtained from a reality-based 3D survey of the circular cavity, with a solid model of the theoretical spherical projectile that produced the impact (shown at **bottom**, **left** to **right**) and detail of the negative mold extrapolated from the ashlar (indicated by the arrow). Geometric analysis by S.B.

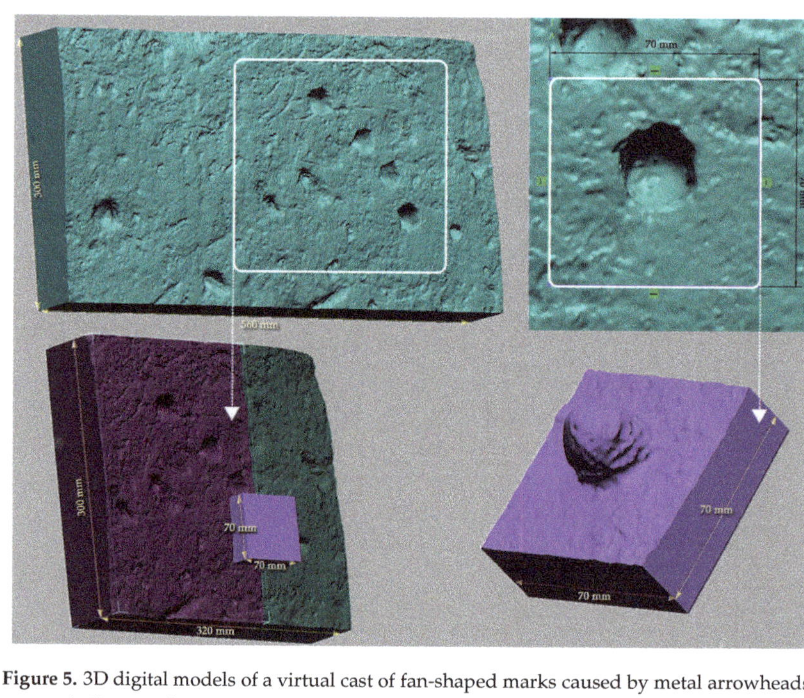

Figure 5. 3D digital models of a virtual cast of fan-shaped marks caused by metal arrowheads. The arrows indicate a close-up of the selected impact holes (left) and the depression from which the negative mold was extracted (left). 3D models and casts by S.B.

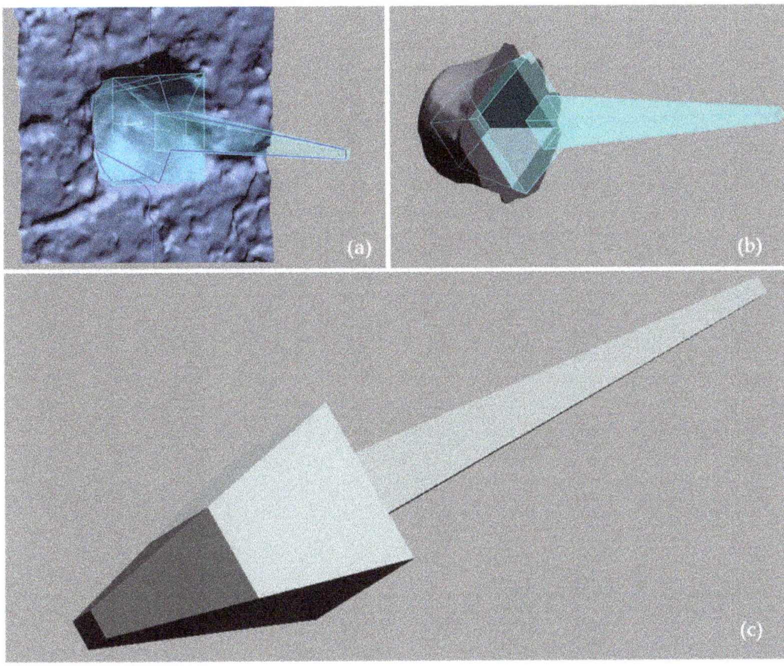

Figure 6. Experimental reconstruction of the metal arrow tip for prototyping, based on the geometry of the existing indentation (**a**,**b**) and the proportions of museum examples (**c**). Models by S.B.

For both examples, the high-resolution polygonal meshes of the impact craters were edited to close any holes and create a support platform. This platform was generated by Boolean operations (union, subtraction, or intersection) between the digital surface and a solid block, appropriately sized for single- or multi-part printing.

Several 3D-printing techniques [21,22] employ additive manufacturing with various materials and levels of detail. From an operational standpoint, the physical reproduction requires a geometry check: meshes must be watertight and manifold before printing. Since raw polygonal surfaces cannot be printed directly, models must be thickened and exported in a suitable format.

In order to prepare each cast model, we defined layer thickness, designed necessary supports, and arranged the nesting on the build plate. The process yielded smoothly finished pieces by printing the STL-converted models at 1:1 scale. We used an Anycubic Predator Fused Deposition Modeling (FDM) printer with PLA (Polylactic Acid) filament, set to a 0.2 mm layer height and 80 mm/s printing speed. Total print times are given in Table 1, and the final prints for both case studies are shown in Figures 7 and 8.

Printing at full scale and allowing later assembly—necessitated by complex overhangs, undercuts, protruding edges, and steep angles (or >45° to the build plate)—required dividing each cast into multiple magnet-joined segments (eight for the cavity cast, four for the mold, and five for the fan-shaped traces). This approach also resolved the challenge of reinserting the negative cast into its positive one, which would have been impossible with rigid material and without release-angle allowances.

Table 1. 3D prints of the case studies shown in Figure 7 (cases a and b) and Figure 8 (case f). 3D prints by C.F.

3D Print Settings	(Figure 7 Case a) Replica of Ballistic Impact	(Figure 7 Case b) Negative Mold	(Figure 8 Case f) Replica of Fan-Shaped Traces
Model scale	1:1	1:1	1:1
Model height	180 mm	180 mm	180 mm
Number of parts	8	4	5
Print layer thickness	0.2 mm	0.2 mm	0.2 mm
Infill percentage	10%	10%	10%
Infill pattern	grid	grid	grid
Estimated print time	50 h	32 h	21 h

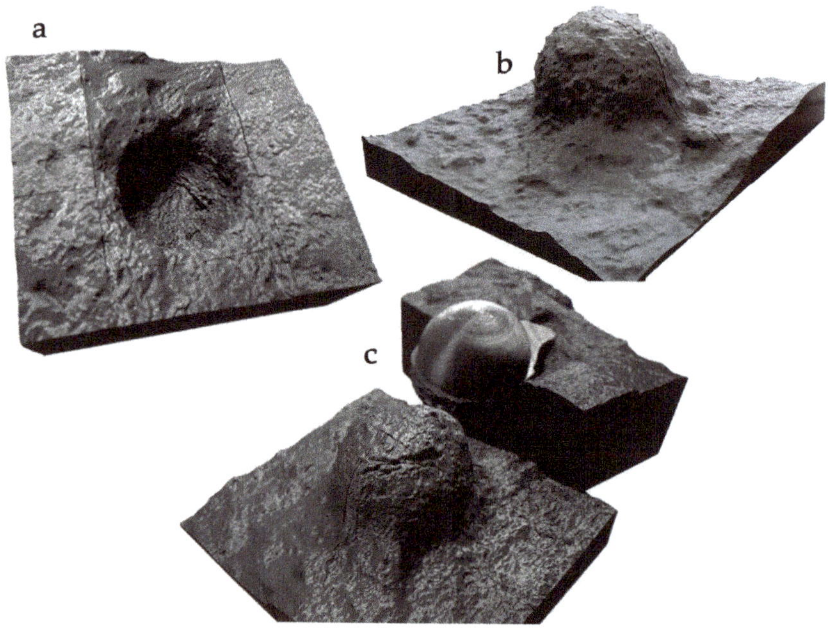

Figure 7. Prototyping of the two cavity types (stone projectiles): (**a**) replica of a ballistic impact based on the digital survey of the city walls; (**b**) negative mold; (**c**) theoretical spherical projectile. 3D-printed prototypes by C.F.

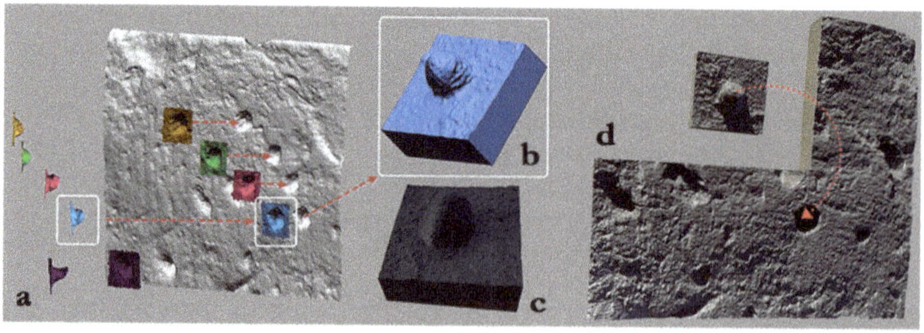

Figure 8. *Cont.*

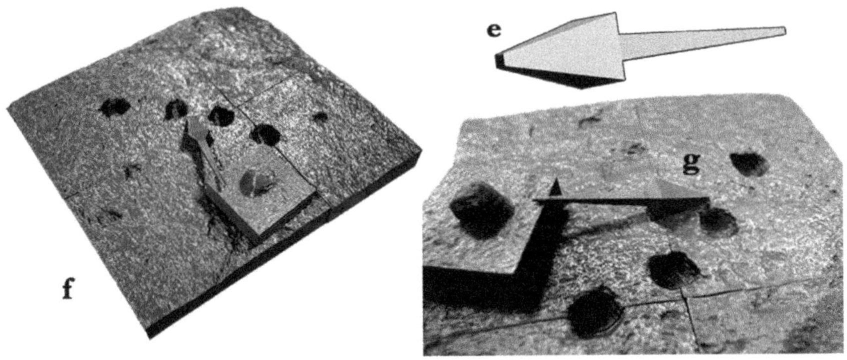

Figure 8. Prototyping of the two cavity types (darts): (**a**) digital survey of the fan-shaped dart marks; (**b**) virtual mold; (**c**,**d**) 3D-printed reproductions; (**e**) hypothetical digital dart model; (**f**,**g**) 3D prints of the fan-shaped dart cavities and metal arrow tip. Models by S.B.; 3D-printed prototypes by C.F.

4. Discussion

The physical 3D print of a digital model, like other representational forms, establishes an analogy with the original artifact despite being produced in different materials. As with its physical analog, the digital model records reality comprehensively through acquisition technologies (Range-Based Modeling (RBM) and Image-Based Modeling (IBM)) and modeling reconstruction techniques such as BIM—HBIM (Building Information Modeling). For practitioners versed in these tools, this approach yields a multifunctional artifact that integrates knowledge within a data-model ecosystem and supports subsequent heritage interventions.

The reliability of 3D models is certainly crucial. The optimization of workflows [23–25], driven by ongoing hardware and software developments, remains a topic of debate. Standardizing common formats for widely used applications [26], including photogrammetric ones based on SfM/MVS [27], seems more like a necessity than an opportunity for the scientific community involved in the documentation of historical and artistic heritage.

In this broad field of application of new technologies for cultural heritage, photogrammetry software based on computer vision processes has permeated multiple scales of documentation, providing effective solutions to diverse problems [28–30]. Integration into established pipelines has led to significant advancements in documentation, even in the architectural domain [31–38].

The digital model produced by integrating diverse acquisition methods provides the foundation for an artifact–knowledge pathway, within which rapid prototyping condenses key innovations. These advances lead to the development of new sectors in geometric modeling software as well as applications aimed at preparing 3D-printable files [39]. Like all additive technologies, fused deposition modeling works in layers, with the minimum layer thickness depending on various factors such as the machine's kinematics, the extrusion nozzle size, and the material used. Final prints must address surface imperfections, tolerance errors, anisotropy from layer orientation, and assembly challenges due to undercuts.

In the context of rapid development, limitations and opportunities must be carefully weighed. This is especially pertinent to recent advances in combining different representation techniques, such as the relation between traditional mesh surface representations and voxels. Voxels are novel units of a representation technique that associate volumetric pixels with the positions of a spatial grid. These volumetric pixels are numerical representations of volume and can be linked to properties such as color, opacity, material density, refractive index, position, force, time, velocity, and more.

As can be inferred, the voxel is a flexibly manipulable element that 3D printing can visualize to some extent by rendering the spatial grid in three dimensions. However, ongoing efforts aim to go further by enabling the accurate printing of voxels along with all their associated properties, to create increasingly complex materials with intrinsic characteristics—an approach often referred to as 4D printing.

Remaining within the field of 3D printing, the experiments presented—like any additive technology—highlight the challenges associated with printing carved or bas-relief works, namely, the indentations on the rough outer surface where the walls form a dihedral angle smaller than a right angle, known as undercuts.

5. Conclusions

The process makes use of digital methodologies that have now become established practices. The need for an acquisition strategy capable of ensuring the resolution required to faithfully reproduce rough cavities—sometimes consisting of holes just a few centimeters deep—represents a case study of interest, and paves the way for future developments. Central to the unified process, which links data acquisition, analysis, and dissemination within a single workflow, is the question this article seeks to address. The case study serves as a means to reflect on the cultural value of prototyping digital casts, highlighting how the printing of 3D models represents a significant mode of operation within a unique, cyclical, and interactive process that connects research, education, and the Third Mission.

Author Contributions: Conceptualization, A.R.; methodology, all authors; software, S.B.; formal analysis and investigation, A.R.; writing—original draft preparation, A.R.; writing—review and editing, S.B.; supervision, A.R.; funding acquisition, A.R.; photogrammetric survey: A.R., S.B. and other members of the UniVanvitelli Team; laser scanner survey, meshing and 3D model elaboration: S.B.; 3D modeling and geometric analysis, S.B.; 3D prints, C.F. Material preparation was performed with different software, in particular Agisoft Metashape (S.B.); Leica Geosystems Cyclone TLS cloud processing and registration (S.B.); 3D Systems Geomagic Design X meshing and processing of polygonal mesh (S.B.); Adobe Photoshop CC elaboration of figures (S.B.). All authors have read and agreed to the published version of the manuscript.

Funding: Claudio Formicola was supported by a scholarship co-financed by DM 118/23 PNRR (M4C1—Inv. 3.4 TDA). This research was supported by the project "SCORPiò-NIDI", CUP B53D23022100006 (DD n. 1012/2023), funded by the Italian Ministry of Research under the PRIN (call DD n. 104/2022) funding initiative.

Institutional Review Board Statement: Not applicable.

Informed Consent Statement: Not applicable.

Data Availability Statement: Research data are available upon request.

Acknowledgments: The authors would like to express their gratitude to the management and the appointed officials of the offices of the Pompeii Archaeological Park for granting authorizations for site access and survey operations.

Conflicts of Interest: The authors declare no conflicts of interest.

Abbreviations

The following abbreviations are used in this manuscript:

3D	Three-Dimensional
BIM/HBIM	Building Information Modeling/Heritage Building Information Modeling
CDE	Common Data Environment
FEA	Finite Element Analysis

FDM	Fused Deposition Modeling
FEM	Finite Element Method
IBM	Range-Based Modeling
ICT	Information and Communications Technology
MR	Mixed Reality
RM	Reverse Modeling
PLA	Polylactic Acid
RBM	Range-Based Modeling
SfM	Structure from Motion
STL	Stereo Lithography or Standard Tessellation Language
XR	Extended Reality

References

1. Rossi, A. The Survey of the Ballistic Imprints for a Renewed Image of Unearthed Pompeii. *Nexus Netw. J.* **2024**, *26*, 307–324. [CrossRef]
2. Vitucci, G. (Ed.) *La Guerra Giudaica di Giuseppe Flavio*; Mondadori: Milano, Italy, 1989; ISBN 88-04-32627-1.
3. Russo, F.; Russo, F. *89 A.C. Assedio a Pompei: La Dinamica e le Tecnologie Belliche Della Conquista Sillana di Pompei*; Edizioni Flavius: Pompei, Italy, 2005; ISBN 88-88419-32-2.
4. y García, L.G. *Danni di Guerra a Pompei. Una Dolorosa Vicenda Quasi Dimenticata. Con Numerose Notizie sul «Museo Pompeiano» Distrutto nel 1943*; Studi della Soprintendenza archeologica di Pompei; L'Erma di Bretschneider: Roma, Italy, 2006; ISBN 88-8265-369-2.
5. Russo, F. *L'artiglieria Delle Legioni Romane. Le Macchine da Guerra che Resero Invincibile L'esercito Romano*; Istituto Poligrafico e Zecca dello Stato: Roma, Italy, 2004; ISBN 978-88-240-3444-9.
6. Rossi, A.; Formicola, C.; Barsanti, S.G. Ingegna Romana. Dalle fonti ai modelli, dai reperti alle ricostruzioni. *Diségno* **2024**, *14*, 229–238. [CrossRef]
7. Marsden, E.W. *Greek and Roman Artillery: Technical Treatises*; Clarendon Press: Oxford, UK, 1971; ISBN 978-0-19-814269-0.
8. Brusaporci, S. Architectural Heritage Imaging: When Graphical Science Meets Model Theory. *Disegnarecon* **2023**, *16*, 1–4. [CrossRef]
9. Migliari, R. (Ed.) *Disegno Come Modello. Riflessioni del Disegno nell'età Informatica*; Edizioni Kappa: Roma, Italy, 2004; ISBN 88-7890-605-0.
10. Cipriani, L.; Fantini, F. Digitalization Culture vs Archaeological Visualization: Integration of Pipelines and Open Issues. *Int. Arch. Photogramm. Remote Sens. Spat. Inf. Sci.* **2017**, *XLII-2-W3*, 195–202. [CrossRef]
11. Sdegno, A.; Cabezos-Bernal, P.M. Oblique Analog Models. *Diségno* **2024**, *14*, 7–21. [CrossRef]
12. Manetti, A. *Vita di Filippo Brunelleschi. Preceduta da la Novella del Grasso*; De Robertis, D., Tanturli, G., Eds.; Edizioni Il Polifilo: Milano, Italy, 1976.
13. Filarete. *Trattato di Architettura*; Grassi, L., Ed.; Edizioni Il Polifilo: Milano, Italy, 1972.
14. Alberti, L.B. *L'architettura (De re Aedificatoria)*; Edizioni Il Polifilo: Milano, Italy, 1966; ISBN 978-88-7050-101-8.
15. Millon, H.; Munshower, S.S. (Eds.) *An Architectural Progress in the Renaissance and Baroque: Sojourns In and Out of Italy*, 1st ed.; Penn State Department of Art History: University Park, PA, USA, 2002; ISBN 978-0-915773-07-7.
16. Millon, H.A.; Smyth, C.H. *Michelangelo Architetto. La facciata di San Lorenzo e la cupola di San Pietro*; Olivetti: Milano, Italy, 1988.
17. Millon, H.A. I modelli architettonici nel Rinascimento. In *Rinascimento: Da Brunelleschi a Michelangelo. La Rappresentazione Dell'architettura*; Millon, H.A., Magnago Lampugnani, V., Eds.; Bompiani: Milano, Italy, 1994; pp. 19–72, ISBN 88-452-2266-7.
18. De Luca, V.; Gatto, C.; Liaci, S.; Corchia, L.; Chiarello, S.; Faggiano, F.; Sumerano, G.; De Paolis, L.T. Virtual Reality and Spatial Augmented Reality for Social Inclusion: The "Includiamoci" Project. *Information* **2023**, *14*, 38. [CrossRef]
19. Calandriello, A.; D'Acunto, G.; Gigliotti, G.C. Tactile Translations: Algorithmic Modelling for Museum Inclusiveness. In *Graphic Horizons*; Hermida González, L., Xavier, J.P., Amado Lorenzo, A., Fernández-Álvarez, Á.J., Eds.; Springer Nature: Cham, Switzerland, 2024; pp. 308–315.
20. Nigro, L.; Montanari, D.; Sabatini, S.; De Giuseppe, M.; Benedettucci, F.M.; Lucibello, S.; Fattore, L.; Trebbi, L.; Nejat, B.; Rinaldi, T. Caress the pharaoh. The tactile reproduction of Ramses II's "mummy" in the Sapienza University Museum of the Near East, Egypt and Mediterranean. *J. Cult. Herit.* **2024**, *67*, 158–163. [CrossRef]
21. Sable, U.; Borlepwar, P.T. Recent Developments in the Field of Rapid Prototyping: An Overview. In *Proceedings of the International Conference on Intelligent Manufacturing and Automation*; Vasudevan, H., Kottur, V.K.N., Raina, A.A., Eds.; Springer: Singapore, 2019; pp. 511–519.

22. Kantaros, A.; Ganetsos, T.; Petrescu, F.I.T. Three-Dimensional Printing and 3D Scanning: Emerging Technologies Exhibiting High Potential in the Field of Cultural Heritage. *Appl. Sci.* **2023**, *13*, 4777. [CrossRef]
23. Guidi, G.; Frischer, B.D. 3D Digitization of Cultural Heritage. In *3D Imaging, Analysis and Applications*; Liu, Y., Pears, N., Rosin, P.L., Huber, P., Eds.; Springer International Publishing: Cham, Switzerland, 2020; pp. 631–697, ISBN 978-3-030-44070-1.
24. Ferdani, D.; Ronchi, D.; Fanini, B.; Manganelli Del Fà, R.; d'Annibale, E.; Bordignon, A.; Pescarin, S. Brancacci Chapel in Florence: Surveying and Real-Time 3D Simulation for Conservation and Communication Purposes. *Int. Arch. Photogramm. Remote Sens. Spat. Inf. Sci.* **2023**, *XLVIII-M-2–2023*, 535–540. [CrossRef]
25. Grasso, N.; Spadavecchia, C.; Di Pietra, V.; Belcore, E. LiDAR and SfM-MVS Integrated Approach to Build a Highly Detailed 3D Virtual Model of Urban Areas. In Proceedings of the 9th International Conference on Geographical Information Systems Theory, Applications and Management; SCITEPRESS—Science and Technology Publications: Prague, Czech Republic, 2023; pp. 128–135.
26. Foschi, R.; Fallavollita, F.; Apollonio, F.I. Quantifying Uncertainty in Hypothetical 3D Reconstruction-A User-Independent Methodology for the Calculation of Average Uncertainty. *Heritage* **2024**, *7*, 4440–4454. [CrossRef]
27. Attenni, M.; Bartolomei, C.; Inglese, C.; Ippolito, A.; Morganti, C.; Predari, G. Low cost survey and heritage value. *SCIRES-IT—Sci. Res. Inf. Technol.* **2018**, *7*, 115–132. [CrossRef]
28. Gaiani, M.; Apollonio, F.I.; Fantini, F. Evaluating Smartphones Color Fidelity and Metric Accuracy for the 3D Documentation of Small Artifacts. *Int. Arch. Photogramm. Remote Sens. Spat. Inf. Sci.* **2019**, *XLII-2-W11*, 539–547. [CrossRef]
29. Apollonio, F.I.; Fantini, F.; Garagnani, S.; Gaiani, M. A Photogrammetry-Based Workflow for the Accurate 3D Construction and Visualization of Museums Assets. *Remote Sens.* **2021**, *13*, 486. [CrossRef]
30. Peinado-Santana, S.; Hernández-Lamas, P.; Bernabéu-Larena, J.; Cabau-Anchuelo, B.; Martín-Caro, J.A. Public Works Heritage 3D Model Digitisation, Optimisation and Dissemination with Free and Open-Source Software and Platforms and Low-Cost Tools. *Sustainability* **2021**, *13*, 13020. [CrossRef]
31. Bevilacqua, M.G.; Russo, M.; Giordano, A.; Spallone, R. 3D Reconstruction, Digital Twinning, and Virtual Reality: Architectural Heritage Applications. In Proceedings of the 2022 IEEE Conference on Virtual Reality and 3D User Interfaces Abstracts and Workshops (VRW), Christchurch, New Zealand, 12–16 March 2022; pp. 92–96.
32. Menaguale, O. Digital Twin and Cultural Heritage—The Future of Society Built on History and Art. In *The Digital Twin*; Crespi, N., Drobot, A.T., Minerva, R., Eds.; Springer International Publishing: Cham, Switzerland, 2023; pp. 1081–1111, ISBN 978-3-031-21343-4.
33. Fascia, R.; Barbieri, F.; Gaspari, F.; Ioli, F.; Pinto, L. From 3D Survey to Digital Reality of a Complex Architecture: A Digital Workflow for Cultural Heritage Promotion. *Int. Arch. Photogramm. Remote Sens. Spat. Inf. Sci.* **2024**, *XLVIII-2-W4-2024*, 205–212. [CrossRef]
34. Grazianova, M.; Mesaros, P. Cultural heritage management from traditional methods to digital systems: A review from bim to digital twin. *E3S Web Conf.* **2024**, *550*, 01015. [CrossRef]
35. Gabellone, F. Digital Twin: A new perspective for cultural heritage management and fruition. *Acta IMEKO* **2022**, *11*, 1–7. [CrossRef]
36. Grieves, M.W. Digital Twins: Past, Present, and Future. In *The Digital Twin*; Crespi, N., Drobot, A.T., Minerva, R., Eds.; Springer International Publishing: Cham, Switzerland, 2023; pp. 97–121, ISBN 978-3-031-21343-4.
37. Luther, W.; Baloian, N.; Biella, D.; Sacher, D. Digital Twins and Enabling Technologies in Museums and Cultural Heritage: An Overview. *Sensors* **2023**, *23*, 1583. [CrossRef] [PubMed]
38. Mousavi, Y.; Gharineiat, Z.; Karimi, A.A.; McDougall, K.; Rossi, A.; Gonizzi Barsanti, S. Digital Twin Technology in Built Environment: A Review of Applications, Capabilities and Challenges. *Smart Cities* **2024**, *7*, 2594–2615. [CrossRef]
39. Gonizzi Barsanti, S.; Marini, M.R.; Malatesta, S.G.; Rossi, A. Evaluation of Denoising and Voxelization Algorithms on 3D Point Clouds. *Remote Sens.* **2024**, *16*, 2632. [CrossRef]

engineering proceedings

MDPI

Proceeding Paper

Denoising and Voxelization for Finite Element Analysis: A Review †

Sara Gonizzi Barsanti

Department of Engineering, Università degli Studi della Campania Luigi Vanvitelli, Via Roma 29, 81031 Aversa, Italy; sara.gonizzibarsanti@unicampania.it

† Presented at the Conference "Discovering Pompeii: From Effects to Causes—From Surveying to the Reconstructions of Ballistae and Scorpiones", Aversa, Italy, 27 February 2025.

Abstract: The conservation of cultural heritage is fundamental, and it is difficult to predict how heritage objects will relate with structural damages. For these objects, the most used process for the analyses involves NURBS models that may introduce an excessive level of approximation leading to wrong simulation results. This work presents the preliminary review of literature and first tests regarding denoising and voxel algorithms and their application for the creation of volumetric models of a reconstruction of an ancient scorpionide, to identify the bottlenecks of the post-processing method for the creation of volumetric data for the FEA of cultural heritage.

Keywords: retopology; denoising algorithms; voxel; structural analysis; geometric analysis; FEA

1. Introduction

World cultural heritage is jeopardized by hazards, both natural (e.g., floods, earthquakes, fires) and man-made (e.g., pollution, mass rapid tourism, traffic, urban sprawl, neglect). Lack of money, interest, and conflicts can also increase the loss of this patrimony. Cultural heritage sites face significant threats during emergencies, recovery phases, and reconstruction efforts following environmental calamities like earthquakes and floods. This is because reconstruction projects, if not carefully planned, can pose a serious danger to historically and culturally significant areas. Consequently, it is crucial to implement risk mitigation strategies to safeguard both movable and immovable cultural heritage, along with their encompassing landscapes. The conservators and owners of cultural sites and museums need a simplified risk analysis approach that does not require extensive expertise to implement. This will allow them to calibrate and organize the conservation process in the best way. It is also an important resource for decision-makers who may not have sufficient knowledge and skills to deal with the complex risk assessment and evaluation process [1]. Diagnostic studies are fundamental for the conservation of cultural heritage; therefore, it is necessary to select and appropriately use current remote non-destructive testing (NDT) techniques. These techniques allow for examining different types of damage on a range of different materials without taking samples from or touching the artifacts. The use of NDT allows for diagnostics that assist in inspection and conservation. Three-dimensional reality-based modeling is an established technique that allows for obtaining 3D models of the state-of-the-art of the object surveyed with high accuracy and precision [2]. The models can be used directly for structural analysis with a strong simplification of the surface elements [3], which allows for maintaining a good accuracy. However, this involves several passages and an increasing level of possible errors or approximation. A good option is to use voxels directly on 3D point clouds, avoiding all the passages. In this paper, the analysis of the application of voxels on a clean point cloud using denoising algorithms is described using a copy of a Xanten–Wardt I sec AD scorpionide as the test object, which is reconstructed by Flavio Russo for Archeotecnica.it in scale 1:1 [4]. The small dart launcher was discovered in 1999 on the bottom of a river, known today as Lake Sudsee in the government district of Düsseldorf.

1.1. Denoising

Three-dimensional point clouds generated from photogrammetric and laser scanning surveys are most of the time noise due to non-collaborative materials or surfaces, inadequate lighting, too complex or convoluted geometries, and the imprecision of the tools employed. This noise not only creates a set of point clouds that are not compliant with the geometry of the object surveyed, but also introduces erratic information, thereby diminishing the geometric accuracy of the mesh model resulting from the following steps of the reality-based process; therefore, this leads to the outcomes of any analyses conducted on it. As a result, it is essential to clean the raw data. The noise removal has emerged as a significant focus in 3D geometric data processing. This process aids in eliminating noise to recover the true original shape of the object surveyed. Bilateral filtering [5] is a non-linear technique utilized mainly for image smoothing and then adapted for point cloud denoising [6]. It considers the position, normal, and color of the points [7]. A second algorithm that can be used is the guided filtering [8], an explicit image filter that functions as an edge-preserving smoothing operator [9]. There are also filter-based algorithms that utilize the normals of the points as guiding signals, leading to an iterative filtering process that updates the points to align them with the estimated normals. Then, graph-based point cloud denoising methods interpret the input point cloud as a graph signal, subsequently performing denoising through selected graph filters [10] and patch-based graph methods that construct patches of point clouds, with each patch represented as a node [11]. Optimization-based denoising techniques denoise the point cloud in an effort to maintain the best approximation of the input point cloud [12]. Lastly, deep learning algorithms have been employed in point cloud processing [10], wherein the noisy inputs are used as a tool for learning a mapping that overlays the ground truth data in an offline phase. Deep learning-based approaches can be categorized into two types: unsupervised [10] and supervised denoising methods such as PointNet [13].

1.2. Voxel

Voxelization is used in object detection, especially for autonomous driving or element recognition in the majority of related works [14–19], or 3D point cloud segmentation. The medical field is another scientific area where voxel-based modeling is used extensively [20–24]. Tests of using voxels for FEA have been conducted. For instance, voxel-based micro modeling was used to create a parametrical model of the composite structure to calculate ballistic impacts on ceramic-polymer composite panels [25]. Another study predicted roof collapses using voxel modeling of caves. This method made it possible to overcome the limitations of the FEM software (Ansys 19.2) and the challenges associated with reconstructing the geometry of the caves [26]. Then, to increase the accuracy of FEA, a homogenization approach for the voxel elements is presented in [27,28]. This is because, although employing voxels speeds up the mesh generation process, it lacks accuracy when working with curved surfaces. Voxelization can be carried out automatically with specific software, for example, Blender 4.1 and Meshmixer 3.5.474, starting from the input mesh and using the open-source library, Open3D. Blender 4.1 has the most straightforward process, but the operator can only control the number of the elements approximated. Meshmixer

3.5.474 creates a watertight solid from mesh surfaces by recomputing the object into a voxel representation. The process is easy, the only parameter the operator can change is the solid type as either fast or accurate, and the solid accuracy with a sliding tool gives a certain number and mesh density. These numbers are not correlated to the final number of polygons of the volume [29]. The Open3D core features include (i) 3D data structures, (ii) 3D data processing algorithms, (iii) scene reconstruction, (iv) surface alignment, (v) and 3D visualization.

1.3. Finite Element Analysis (FEA)

The finite element method (FEM) is a numerical technique used to perform finite element analysis (FEA), initially developed for structural mechanics and then applied to the solution of other types of problems [30]. The physical problem usually involves a structure, or a structural component subjected to certain loads. The idealization of this type of problem into a mathematical one involves specific assumptions that lead to differential equations. The finite element analysis solves this mathematical model, and since this solution technique is a numerical procedure, the accuracy of the solution must be considered. If the accuracy is not met, the numerical solution must be repeated with refined parameters (for example, finer meshes) until an adequate accuracy is achieved. The finite element analysis approximates the exact solution of the problem and the behavior of any point within the finite element is described by the nodal displacement, which is the first result of an FEM calculation. In summary, the use of the FEM increases accuracy, improves design, and enhances the identification of critical parts of a structure or object. In FEA, a mathematical model is commonly used as an idealization of the physical object built to predict or simulate its behavior. The analysis is then performed on the meshed models using data elements; this is different if a 2D or 3D problem is evaluated. Two-dimensional elements are triangular and quadrangular: quadrangular elements are preferred since triangular ones have lower precision. Three-dimensional elements are tetrahedral and hexahedral: as for 2D elements, hexahedral ones are more accurate (e.g., they deform to a lower strain energy state), but it is more difficult to mesh a 3D volume with this type of element if it is not segmented. Three-dimensional elements can be linear or quadratic: the difference is that quadratic ones have nodes also in the central side, varying the number from four nodes (linear tetrahedron) to 20 nodes (quadratic hexahedron).

2. Materials and Methods

This research utilized the Scopus and Web of Science (WOS) databases as primary search tools. Scopus, recognized as the largest abstract and citation database globally, encompasses over 20,000 journals across various disciplines published by more than 5000 publishers. Its extensive coverage and focus on specific subjects provide researchers with access to a broader spectrum of literature, thereby offering more robust data support to the academic community. To reduce personal bias and enhance the overall quality of the review, this study implemented a structured data search methodology in accordance with the meta-analysis protocol for systematic reviews. The initial phase of this literature review involves an exploratory survey designed to identify studies that have employed both techniques in examining climate-induced degradation processes impacting historical buildings. The systematic literature review adheres to the PRISMA (Preferred Reporting Items for Systematic Reviews and Meta-Analyses) guidelines [31].

2.1. Exploratory Survey for Keywords Detection

The exploratory survey enables the detection of trusted and specific keyword combinations. This survey was conducted by searching documents that simultaneously utilized the denoising, voxel, and 3D survey for FEA keywords. In Scopus, the process of identifying the relevant documents to be analyzed in detail was conducted by searching within the fields "Article title, Abstract and Keyword" on the Search page. This search included all documents in the database published from 2022 through the end of January 2025. Since the intention was to analyze if denoising and voxel algorithms have been used together and if they have been used in combination with FEA in cultural heritage, the research was performed with 3 keyword combinations using the Boolean operator "AND": (1) "voxel" AND "FEA" AND "point cloud" (27 papers); (2) "voxel" AND "FEA" AND "point cloud" AND "cultural heritage*" (242 papers); (3) "voxel" AND "denoising" AND "cultural heritage*" (5 papers). Initially, the search yielded a total of 265 documents; successively, the group was narrowed down by reviewing the titles and excluding those papers that (i) did not appear related to the simultaneous use of the techniques, (ii) were doubled, and (iii) were either unavailable or written in English. From the screening process, 185 documents were excluded, resulting in a total of 71 papers. The literature review for this contribution focuses on the less-explored domain of the simultaneous application of denoising and voxel algorithms for the structural analysis of cultural heritage and photogrammetry.

2.2. Screening and Inclusion Phases

The second step of the screening process included the title and abstract. Articles "out of topic" were excluded as they dealt with the following topics:

- Type of applications out of topic such as medical or additive manufacturing research.
- Type of applications regarding built environment but not inherent with cultural heritage (both movable and immovable), such as aqueducts, viaducts, bridges and/or very modern structures (built environment <20 years).
- Papers particularly focused on the effectiveness of these techniques separately, without exploring their joint or simultaneous use for monitoring the condition of historical structures or heritage objects.

This step resulted in the exclusion of 51 documents, leading to the final selection of 13 publications, all in scientific journals.

2.3. State-of-the-Art in the Integration of Denoising Algorithms and Voxels

All the selected papers were analyzed based on their research approach. The first group analyzed was the one considering the keywords "voxelANDdenoisingANDculturalheritage". Unfortunately, none of the papers presented a combined use of voxels and denoising algorithms for the post processing of cultural heritage models. Refs. [32–35] considered Deep and Machine Learning algorithms for the classification and representation of 3D point clouds, and three of them are reviews. None of the papers considered denoising algorithms or voxels for the optimization and post processing of point clouds. The use of voxels, analyzed in [36], provides a review of the use of applications using voxel-based representations, while in [37,38], the application of Machine and Deep Learning algorithms for the preservation of cultural heritage is considered, but not the ones investigated in this paper. Finally, in [39], a work on a hybrid survey technique for cultural heritage is presented. In summary, none of the papers identified have considered the two processes together. The second group, "voxelANDfeaANDpointcloud", is the one in which more papers have been selected. In [40], an interesting approach on damage identification is presented, but it is not related to cultural heritage. The topology optimization is analyzed in [41], but again the topic is not cultural heritage and no denoising algorithms or voxels have been analyzed. The use of algorithms for the prediction of cracks is investigated in [42], while [43] presents a work on data structures and algorithms for efficiently converting piece-wise linear geometric data into topologically adequate voxel data. The last group is the one from the keywords "voxel+feaANDpointcloudANDculturalheritage". In [44], the only paper left after the two processes of screening, the FEA for cultural heritage is considered, but not Deep and Machine Learning algorithms. It is hence clear that the field is not characterized by comprehensive works regarding the combined use of these algorithms, and even if the investigation follows just one of the two processing methods, the goal of the research is not FEA.

3. Results

Considering the lack of state-of-the-art, it was decided to test the most contemporary algorithms on a 3D point cloud of a copy of the scorpionide. A point cloud derived from a photogrammetric survey has been processed with denoising and voxel algorithms to analyze the difficulties and the problems in applying these algorithms to complex structures. It was decided to use the copy of a Roman scorpionide, a war throwing machine used in sieges (Figure 1).

Figure 1. Image of the copy of the Xanten–Wardt scorpionide, reconstructed by Flavio Russo in scale 1:1.

3.1. Denoising

The point cloud is processed using Agisosft Metashape 2.2.1 and scaled using specific metric targets. The algorithm used for point cloud cleaning is one of the most relevant algorithms in recent research [45]. The proposed method consists of a denoiser based on the score of the point cloud in three-dimensional space. It uses deep learning principles to train the model and identify the best fitting region for each point. A gradient method is used to align the outlier groups with the estimated evaluation function. In other words, the technique simulates an intelligent smoothing of the underlying surface based on the majority voting method (or point density/size). The following methodology was employed: (i) use of a Git repository of the starting project [46]; (ii) creation of an Anaconda environment, with an attention to project requirements and packages needed; (iii) large point cloud denoising (>50,000 points) was run on the model; (iv) an XYZ mesh was saved as an output. The system uses two libraries: pytorch3d and pytorch, the first for managing 3D elements and the second for training deep learning models. The latter is the core of the system extracting features from the input, then giving the data to the evaluation network, which classifies them using a threshold-based clustering principle. Then, it aligns the points to match the threshold, and removes the outliers. The original point cloud and the denoised are then compared using CloudCompare software 2.14 giving as result a Gaussian distribution of mean and standard deviation and a signed C2C (cloud-to-cloud) distance, this means that only the closest points are compared (Figure 2).

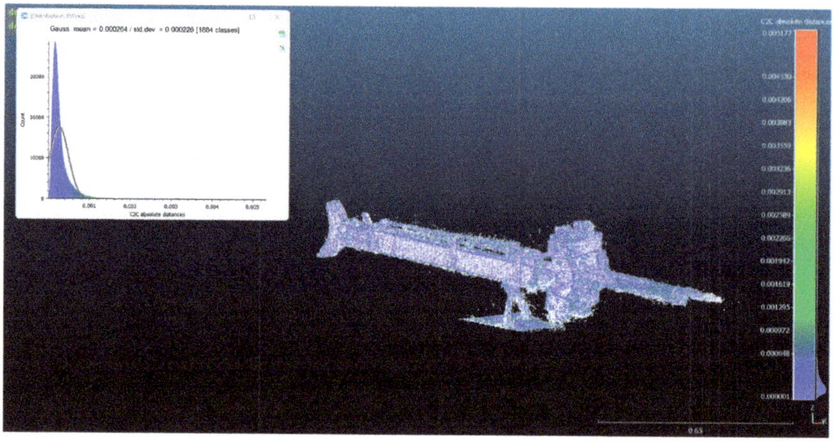

Figure 2. Comparison of the two point clouds.

To better evaluate the differences between the two point clouds, a series of profiles were exported in *.dxf format and analyzed using Rhinoceros software 7 to determine the maximum distance between the two lines (marked red for the non-denoised point cloud and blue for the denoised one). The overlap of these profiles illustrates how the algorithm impacts the geometric organization of the 3D points, realigning noisy data to match the ideal surface. The profiles reveal discrepancies, with the red line deviating significantly from the true geometry of the object (Figure 3a,b).

(**a**)

Figure 3. *Cont.*

(b)

Figure 3. Comparison of profiles of the raw 3D point cloud (red) and the denoised one (blue). (**a**) The entire profile extrapolated with a section of the 3D point cloud; (**b**) the detail of the maximum distance of the two point clouds. It is evident how the denoising algorithm can rearrange the surface, avoiding spikes and data that are geometrically incoherent.

The results are summarized in Table 1.

Table 1. Mean and standard deviation of the scorpionide point cloud analyzed after denoising.

Object	Mean (mm)	Standard Deviation (mm)	Profiles Max Distance (mm)
Scorpionide	0.000264	0.000226	16,764

3.2. Voxelization

The problem when dealing with complex geometries, such as those of cultural heritage artifacts, is that voxelization algorithms present too much simplification. A strong algorithm to process voxels is presented in [46]. Unfortunately, the starting point is a mesh, and it was decided to start from the point cloud to create voxel grids. Therefore, it was decided to test the open source Open3D library [47] with the function voxel_down_sample (self, voxel_size) that allows for downsampling the input point cloud to an output point cloud with one voxel. Normals and colors are averaged if they exist. The steps of the methodology applied are as follows:

- Imposition of voxel size to downsample;
- Definition of the resolution of the voxel grid;
- Creation of the voxel grid (Figure 4);
- Creation of a binary occupancy grid;
- Activation of a corresponding index in the binary grid for each voxel;
- Application of the Marching Cubes algorithm to extract the mesh surface;
- STL file is then saved.

Starting from a simple visual analysis, (Figure 5a,b), the result is promising: the algorithm managed to create a complete model, being able to follow also the most geometrically complex parts, such as the ropes and the small metal nails.

To analyze further the accuracy of the data, the voxel model was then compared to both the denoised point cloud and the mesh model derived (Figure 6a,b).

The results are shown in Table 2.

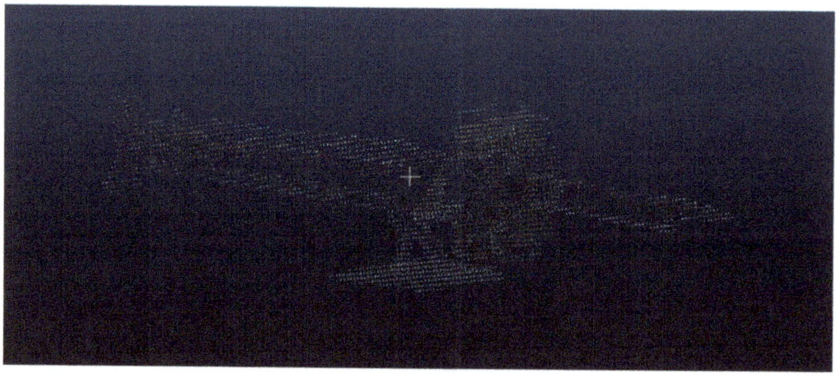

Figure 4. The voxel grid created from the denoised point cloud.

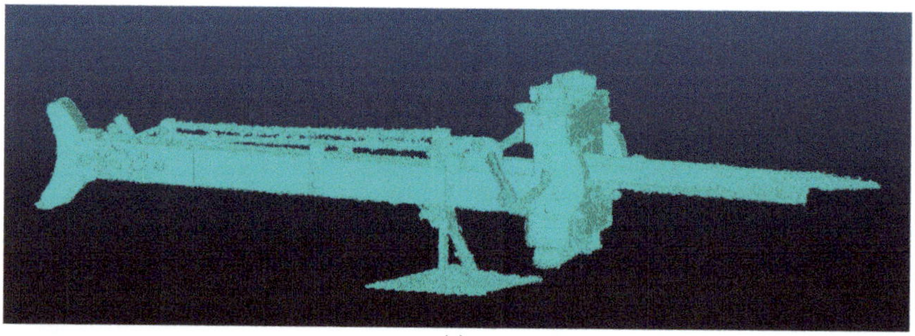

(a)

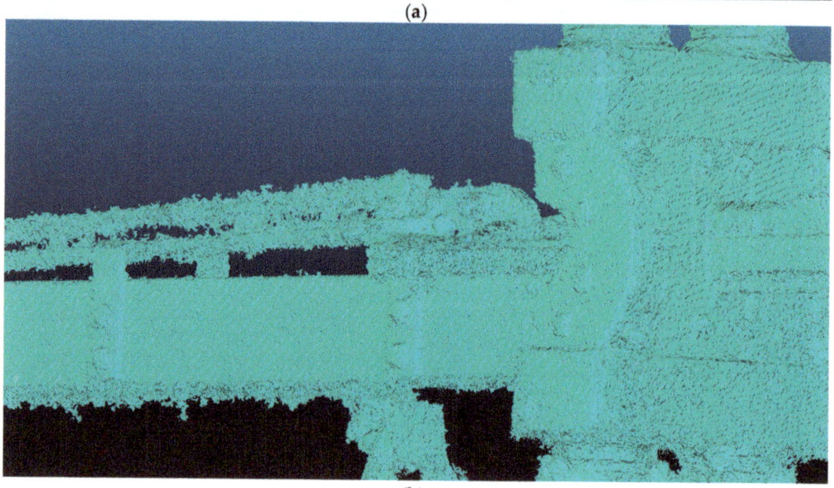

(b)

Figure 5. The voxel model (a) obtained from the denoised point cloud and a zoom of the front part (b).

Table 2. Mean and standard deviation of the voxel model of the scorpionide compared with the denoised point cloud and the mesh.

	Mean (mm)	Standard Deviation (mm)
Cfr denoised/voxel (mm)	0.00008	0.000393
Cfr voxel/mesh	0.000369	0.000760

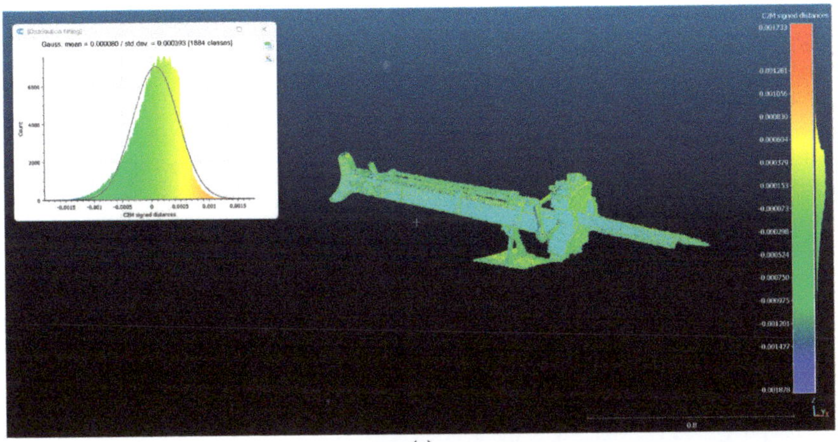

(**a**)

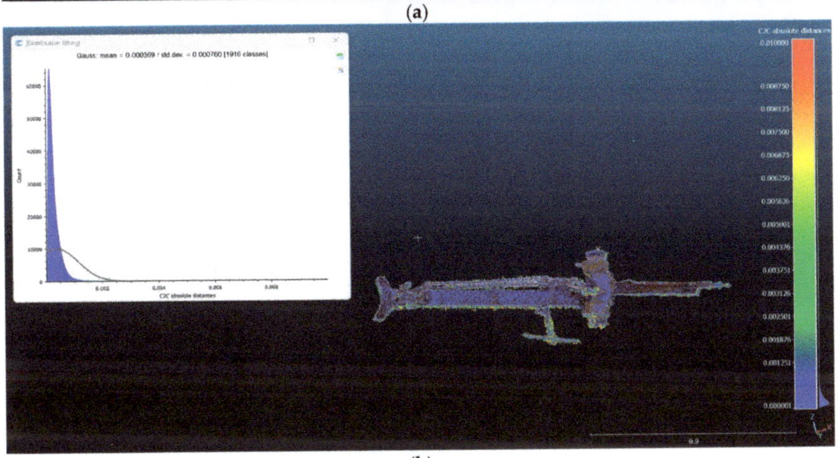

(**b**)

Figure 6. The voxel model obtained compared with the denoised point cloud (**a**) and the mesh (**b**).

4. Conclusions

FEA can work with 2D or 3D models, depending on the object to be analyzed. Volumetric models usually employed in the structural analysis software are formed by NURBS patches through 3D CAD modeling. If the analysis is carried out on models of cultural heritage, one possibility is to directly use 3D reality-based models transformed in volumes. This procedure, even if proved to be effective [3], consists of a sequence of iterative steps, each one contributing to add a degree of approximation to the final result. Starting from a reality-based survey, the data, after the conversion of a disorganized point cloud into a mesh via 3D surface reconstruction (first level of approximation) can be redrawn with CAD software from profiles, and then exported in BIM or HBIM models or converted to Non-Uniform Rational B-Splines (NURBS) following the process that uses retopology for the simplification of the mesh [45]. The export of 3D superficial meshes into NURBS creates one patch for each element of the surface mesh, thereby approximating the shape of the object. Additionally, the application of finite element analysis (FEA) in structural analysis introduces another step of approximation, leading to a cumulative effect that can cause the final output to significantly diverge from reality. Consequently, starting with less precise data, such as a point cloud that has a large amount of noise due to survey issues, inevitably results in a less accurate final product.

The denoising algorithm is effective because it allows for improving geometric accuracy, which means a more accurate and precise distribution of the points. Therefore, leading to a stronger closeness of the geometry of the point cloud to the geometry of the real object surveyed.

To skip a few passages and to decrease the level of approximation of the volume used in FEA, voxelization algorithms can be used to directly transform 3D point clouds in volume. The tessellated surface can be a problem while analyzing the model in FEA software, especially during the meshing process, and it is a problem that still needs to be fixed. In this paper, the use of precompiled tools and available library for the creation of volumes from 3D reality-based models of the object have been used starting from a first attempt of literature review, with just three searching combinations of keywords. It was decided to narrow the search on these three specific combinations to have an immediate idea of the state-of-the-art.

The test of the Open3D library gave optimal results in the creation of the voxel grid from 3D point clouds. This is why the denoising part has to be considered as best practice and the first step for obtaining a better voxel grid. The use of voxels for the creation of volumes shows great results in the medical field, probably because the data used have a different accuracy, geometry, and dimensions. For FEA, on the other hand, some problems arose during the test because the geometry of the volume, even if good in terms of standard deviation compared to the input data, is not well-read by the FEA software. Few tests have been carried out and not presented in this paper since the results are poor, but the main issues are regarding the meshing step with 3D elements and the imposition of forces and boundary conditions. One possibility for the latter is to segment the point cloud and then create a volume for each segmented part. This step may occur, however, in problems related to the continuity of the adjacent surfaces that shape the model. If these volumetric surfaces are not strictly connected, the FEA will give inaccurate or wrong results. This process is the first of a series of difficult improvements that the available tools for voxelization need. The reason lies, probably, in the different fields in which voxels have always been used, fields that do not need this level of accuracy or deal with data that are geometrically less complex.

The intention for future works is to analyze different algorithms and software for voxel creation using denoised point clouds of objects that vary in geometry, composition of materials, and dimensions. What is at least expected is to find, hopefully, the bottleneck of the available algorithms that can be improved for a more accurate result. From the first stage of state-of-the-art review, currently there is no script or algorithm nor software that is able to provide the level of accuracy needed for the pipeline proposed.

Funding: This research was supported by the project "SCORPiò-NIDI", CUP B53D2302210 0006, funded by the Italian Ministry of Research under the PRIN (DD n.104/2022) funding initiative, PI prof. Rosa De Finis, University of Salento, co-PI prof. Sara Gonizzi Barsanti, University of Campania Luigi Vanvitelli.

Institutional Review Board Statement: Not applicable.

Informed Consent Statement: Not applicable.

Data Availability Statement: Research data are available upon request.

Acknowledgments: The author would like to thank Gabriel Zuchtriegel, Director of the Archaeological Area of Pompeii, Giuseppe Scarpati, Head of the Study and Research Area, and Valeria Amoretti. The research activities are part of the MUR–PRIN 2022 project "SCORPiò-NIDI".

Conflicts of Interest: The author declares no conflicts of interest. The funders had no role in the design of the study; in the collection, analyses, or interpretation of data; in the writing of the manuscript; or in the decision to publish the results.

References

1. Pedersoli, J.L., Jr.; Antomarchi, C.; Michalski, S. *A Guide to Risk Management of Cultural Heritage*; ICCROM ATHAR Regional Conservation Centre: Sharjah, United Arab Emirates, 2016.
2. Rossi, A.; Cipriani, L.; Cabezos-Bernal, P.M. 3D Digital Models. Accessibility and Inclusive Fruition. *DISEGNARECON* **2024**, *17*, 1–6. [CrossRef]
3. Gonizzi Barsanti, S.; Guagliano, M.; Rossi, A. 3D Reality-Based Survey and Retopology for Structural Analysis of Cultural Heritage. *Sensors* **2022**, *22*, 9593. [CrossRef] [PubMed]
4. Fratino, M.; Rossi, A. Re-construction of the small Xanten dart launcher. In *Discovering Pompeii: From Effects to Causes. From Surveying to the Reconstructions of Ballistae and Scorpiones*; Real Casa dell'Annunziata, Department of Engineering Vanvitelli University; MDPI: Basel, Switzerland, 2025; *under process of publication*.
5. Tomasi, C.; Manduchi, R. Bilateral filtering for gray and color images. In Proceedings of the Sixth International Conference on Computer Vision, Bombay, India, 7 January 1998; IEEE: Piscataway, NJ, USA, 1998; pp. 839–846.
6. Chen, H.; Shen, J. Denoising of point cloud data for computer-aided design, engineering, and manufacturing. *Eng. Comput.* **2018**, *34*, 523–541. [CrossRef]
7. Digne, J.; Franchis, C.D. The bilateral filter for point clouds. *Image Process. Online* **2017**, *7*, 278–287. [CrossRef]
8. He, K.; Sun, J.; Tang, X. Guided image filtering. *IEEE Trans. Pattern Anal. Mach. Intell.* **2013**, *35*, 1397–1409. [CrossRef]
9. Han, X.; Jin, J.S.; Wang, M.; Jiang, W. Guided 3D point cloud filtering. *Multimed. Tools Appl.* **2018**, *77*, 17397–17411. [CrossRef]
10. Irfan, M.A.; Magli, E. Exploiting color for graph-based 3d point cloud denoising. *J. Vis. Commun. Image Represent.* **2021**, *75*, 103027. [CrossRef]
11. Dinesh, C.; Cheung, G.; Bajić, I.V. Point cloud denoising via feature graph Laplacian regularization. *IEEE Trans. Image Process.* **2020**, *29*, 4143–4158. [CrossRef]
12. Xu, Z.; Foi, A. Anisotropic denoising of 3D point clouds by aggregation of multiple surface-adaptive estimates. *IEEE Trans. Vis. Comput. Graph.* **2021**, *27*, 2851–2868. [CrossRef]
13. Rakotosaona, M.J.; La Barbera, V.; Guerrero, P.; Mitra, N.J.; Ovsjanikov, M. Pointcleannet: Learning to denoise and remove outliers from dense point clouds. *Comput. Graph. Forum* **2021**, *39*, 185–203. [CrossRef]
14. Sun, J.; Ji, Y.M.; Wu, F.; Zhang, C.; Sun, Y. Semantic-aware 3D-voxel CenterNet for point cloud object detection. *Comput. Electr. Eng.* **2022**, *98*, 107677. [CrossRef]
15. He, C.; Li, R.; Li, S.; Zhang, L. Voxel set transformer: A set-to-set approach to 3D object detection from point clouds. In Proceedings of the IEEE/CVF Conference on Computer Vision and Pattern Recognition, New Orleans, LA, USA, 18–24 June 2022; pp. 8417–8427.
16. Mahmoud, A.; Hu, J.S.; Waslander, S.L. Dense voxel fusion for 3D object detection. In Proceedings of the IEEE/CVF Winter Conference on Applications of Computer Vision and Pattern Recognition, Vancouver, BC, Canada, 17–24 June 2023; pp. 663–672.
17. Shrout, O.; Ben-Shabat, Y.; Tal, A. GraVoS: Voxel Selection for 3D Point-Cloud Detection. In Proceedings of the IEEE/CVF Winter Conference on Applications of Computer Vision and Pattern Recognition, Vancouver, BC, Canada, 17–24 June 2023; pp. 21684–21693.
18. Deng, J.; Shi, S.; Li, P.; Zhou, W.; Zhang, Y.; Li, H. Voxel r-cnn: Towards high performance voxel-based 3D object detection. *Proc. AAAI Conf. Artif. Intell.* **2021**, *35*, 1201–1209. [CrossRef]
19. He, C.; Zeng, H.; Huang, J.; Hua, X.S.; Zhang, L. Structure Aware Single-Stage 3D Object Detection from Point Cloud. In Proceedings of the IEEE/CVF Conference on Computer Vision and Pattern Recognition (CVPR), Seattle, WA, USA, 13–19 June 2020; pp. 11870–11879.
20. Lv, C.; Lin, W.; Zhao, B. Voxel Structure-Based Mesh Reconstruction From a 3D Point Cloud. *IEEE Trans. Multimed.* **2022**, *24*, 1815–1829. [CrossRef]
21. Sas, A.; Ohs, N.; Tanck, E.; van Lenthe, G.H. Nonlinear voxel-based finite element model for strength assessment of healthy and metastatic proximal femurs. *Bone Rep.* **2020**, *12*, 100263. [CrossRef] [PubMed]
22. Lee, T.Y.; Weng, T.L.; Lin, C.H.; Sun, Y.N. Interactive voxel surface rendering in medical applications. *Comput. Med. Imaging Graph.* **1999**, *23*, 193–200. [CrossRef]
23. Han, G.; Li, J.; Wang, S.; Wang, L.; Zhou, Y.; Liu, Y. A comparison of voxel- and surface-based cone-beam computed tomography mandibular superimposition in adult orthodontic patients. *J. Int. Med. Res.* **2021**, *49*, 0300060520982708. [CrossRef] [PubMed]
24. Goto, M.; Abe, O.; Hagiwara, A.; Fujita, S.; Kamagata, K.; Hori, M.; Aoki, S.; Osada, T.; Konishi, S.; Masutani, Y.; et al. Advantages of Using Both Voxel- and Surface-based Morphometry in Cortical Morphology Analysis: A Review of Various Applications, Magnetic Resonance. *Med. Sci.* **2022**, *21*, 41–57. [CrossRef] [PubMed]

25. Babich, M.; Kublanov, V. Voxel Based Finite Element Method Modelling Framework for Electrical Stimulation Applications Using Open-Source Software. In Proceedings of the Ural Symposium on Biomedical Engineering, Radioelectronics and Information Technology (USBEREIT), Yekaterinburg, Russia, 25–26 April 2019; pp. 127–130.
26. Sapozhnikov, S.B.; Shchurova, E.I. Voxel and Finite Element Analysis Models for Ballistic Impact on Ceramic-polymer Composite Panels. *Procedia Eng.* **2017**, *206*, 182–187. [CrossRef]
27. Doğan, S.; Güllü, H. Multiple methods for voxel modeling and finite element analysis for man-made caves in soft rock of Gaziantep. *Bull. Eng. Geol. Environ.* **2022**, *81*, 23. [CrossRef]
28. Watanabe, K.; Iijima, Y.; Kawano, K.; Igarashi, H. Voxel Based Finite Element Method Using Homogenization. *IEEE Trans. Magn.* **2012**, *48*, 543–546. [CrossRef]
29. Gonizzi Barsanti, S.; Marini, M.R.; Malatesta, S.G.; Rossi, A. Evaluation of Denoising and Voxelization Algorithms on 3D Point Clouds. *Remote Sens.* **2024**, *16*, 2632. [CrossRef]
30. Zienkiewicz, O.C.; Taylor, R.L. *The Finite Element Method*; McGraw Hill: London, UK, 1989.
31. Page, M.J.; McKenzie, J.E.; Bossuyt, P.M.; Boutron, I.; Hoffmann, T.C.; Mulrow, C.D.; Shamseer, L.; Tetzlaff, J.M.; Akl, E.A.; Brennan, S.E.; et al. The PRISMA 2020 statement: An updated guideline for reporting systematic reviews. *BMJ* **2021**, *372*, n71. [CrossRef] [PubMed]
32. Muzahid, A.A.M.; Han, H.; Zhang, Y.; Dawei, L.; Zhang, Y.; Jamshid, J. Ferdous Sohel, Deep learning for 3D object recognition: A survey. *Neurocomputing* **2024**, *608*, 128436. [CrossRef]
33. Xue, F.; Lu, W.; Webster, C.J.; Chen, K. A derivative-free optimization-based approach for detecting architectural symmetries from 3D point clouds. *ISPRS J. Photogramm. Remote Sens.* **2019**, *148*, 32–40. [CrossRef]
34. Shahab, S.S.; Himeur, Y.; Kheddar, H.; Amira, A.; Fadli, F.; Atalla, S.; Copiaco, A.; Mansoor, W. Advancing 3D point cloud understanding through deep transfer learning: A comprehensive survey. *Inf. Fusion* **2025**, *113*, 102601.
35. Di Angelo, L.; Di Stefano, P.; Guardiani, E. A review of computer-based methods for classification and reconstruction of 3D high-density scanned archaeological pottery. *J. Cult. Herit.* **2022**, *56*, 10–24. [CrossRef]
36. Xu, Y.; Tong, X.; Stilla, U. Voxel-based representation of 3D point clouds: Methods, applications, and its potential use in the construction industry. *Autom. Constr.* **2021**, *126*, 103675. [CrossRef]
37. Zu, X.; Gao, C.; Liu, Y.; Zhao, Z.; Hou, R.; Wang, Y. Machine intelligence for interpretation and preservation of built heritage. *Autom. Constr.* **2025**, *172*, 106055. [CrossRef]
38. Liu, D.; Cao, K.; Tang, Y.; Zhang, J.; Meng, X.; Ao, T.; Zhang, H. Study on weathering corrosion characteristics of red sandstone of ancient buildings under the perspective of non-destructive testing. *J. Build. Eng.* **2024**, *85*, 108520. [CrossRef]
39. Wang, Y.; Bi, W.; Liu, X.; Wang, Y. Overcoming single-technology limitations in digital heritage preservation: A study of the LiPhoScan 3D reconstruction model. *Alex. Eng. J.* **2025**, *119*, 518–530. [CrossRef]
40. Gao, Y.; Li, H.; Fu, W.; Chai, C.; Su, T. Damage volumetric assessment and digital twin synchronization based on LiDAR point clouds. *Autom. Constr.* **2024**, *157*, 105168. [CrossRef]
41. Zhang, Z.; Yao, W.; Li, Y.; Zhou, W.; Chen, X. Topology optimization via implicit neural representations. *Comput. Methods Appl. Mech. Eng.* **2023**, *411*, 116052. [CrossRef]
42. Zhao, Y.; Liu, Y.; Xu, Z. Statistical learning prediction of fatigue crack growth via path slicing and re-weighting. *Theor. Appl. Mech. Lett.* **2023**, *13*, 100477. [CrossRef]
43. Nourian, P.; Azadi, S. Voxel graph operators: Topological voxelization, graph generation, and derivation of discrete differential operators from voxel complexes. *Adv. Eng. Softw.* **2024**, *196*, 103722. [CrossRef]
44. Cakir, F.; Kucuk, S. A case study on the restoration of a three-story historical structure based on field tests, laboratory tests and finite element analyses. *Structures* **2022**, *44*, 1356–1391. [CrossRef]
45. Shitong, L.; Hu, W. Score-based point cloud denoising. In Proceedings of the IEEE/CVF International Conference on Computer Vision, Montreal, QC, Canada, 10–17 October 2021; pp. 4563–4572.
46. Baert, J. Cuda Voxelizer: A Gpu-Accelerated Mesh Voxelizer. 2017. 4. Available online: https://github.com/Forceflow/cuda_voxelizer (accessed on 5 February 2025).
47. Zhou, Q.Y.; Park, J.; Koltun, V. Open3D: A modern library for 3D data processing. *arXiv* **2018**, arXiv:1801.09847.

engineering proceedings

Proceeding Paper

Structural Integrity Assessment of Pompeii's City Wall Under Roman Artillery Fire: A Finite Element Approach †

Monil Mihirbhai Thakkar, Amir Ardeshiri Lordejani and Mario Guagliano *

Dipartimento di Meccanica, Politecnico di Milano, Via La Masa 1, 20156 Milan, Italy; monilmihirbhai.thakkar@polimi.it (M.M.T.); amir.ardeshiri@polimi.it (A.A.L.)
* Correspondence: mario.guagliano@polimi.it; Tel.: +39-0223998212
† Presented at the Conference "Discovering Pompeii: From Effects to Causes—From Surveying to the Reconstructions of Ballistae and Scorpiones", Aversa, Italy, 27 February 2025.

Abstract: During Sulla's siege of Pompeii in 89 BC projectiles were launched using Roman artillery, leaving visible craters on the fortified walls. The city was later buried by the eruption in 79 AD, preserving both its architectural layout and the damaged wall surfaces, and was excavated in the early 20th century. This study focuses on simulating projectile impacts on Grey Tuff to estimate impact velocities and penetration depths, offering insights into the destructive capability of Roman weapons. Material models are developed, followed by finite element analysis. Mesh convergence, velocity calibration, and angular impact studies are performed for both ballista and dart to better understand impact mechanics and crater formation.

Keywords: finite element analysis; Nocera Tuff; Pompei; ancient Roman artillery; scorpione; ballista; concrete damaged plasticity

1. Introduction

The defensive walls of Pompeii played a crucial role in the city's protection and urban development, although their exact historical significance and chronology are not explicitly detailed in the provided papers. However, we can draw some insights from the available information. The forum at Pompeii underwent significant changes after the earthquake of 62 CE, with a comprehensive post-earthquake plan that included "blocking streets, linking facades, upgrading building materials, and emphasizing the now more prominent northeastern and southeastern entrances" [1]. This suggests that the city's defensive structures were likely modified or reinforced during this period.

Interestingly, while the reported studies do not provide specific information about Pompeii's walls, they offer insights into fortifications in other ancient cities. For instance, Greek cities built impressive fortifications for defense and protection, with examples ranging from Ephesus to Dura Europus [2]. These structures often served multiple purposes beyond mere defense, as seen in the case of Tecolote, Guatemala, where fortifications were part of an integrated polity-wide system of defense and were used as staging grounds for attacks [3]. Although specific details about Pompeii's walls are limited, insights from other ancient cities indicate that such fortifications often played multifaceted roles, including defense and strategic military functions. This multifunctionality likely extended to Pompeii, influencing the choice of construction materials and methods.

The defensive walls of Pompeii were constructed using a variety of materials, primarily consisting of volcanic rocks from the region. The main components were Grey Tuff or Fiano/Nocera Tuff, which are more rigid and resilient than the extensively studied

Neapolitan Yellow Tuff (NYT) [4]. Nocera Tuff is a grey volcanic stone from the Campanian Ignimbrite (CI) [5,6], a widespread pyroclastic deposit that resulted from a massive eruption of the Campi Flegrei volcanic system around 39,000 years ago [7,8]. This ignimbrite, known for its durability and resistance to weathering, was an ideal choice for fortifications. The walls were likely built using a technique similar to that observed in other Roman structures of the time, involving a concrete core made of lime mortar mixed with volcanic ash (pozzolana) and aggregate [8,9]. This combination created a strong and durable material that could withstand seismic activity and long-term settlement. The use of local volcanic materials, such as tuff and scoria, as aggregate would have further enhanced the strength and durability of the walls [8,9]. The presence of zeolites in some of these volcanic materials may have contributed to the long-term strength development of the concrete through pozzolanic reactions [7,10].

During the Roman siege led by Lucius Cornelius Sulla of Pompeii in 89 BC, the city's fort walls were subjected to intense bombardment from Roman artillery, including ballistae and catapults. The siege was part of the Social War, a conflict between Rome and its Italian allies. The northern walls of Pompeii, which faced a plain, were particularly vulnerable and were reinforced by the Pompeians with additional masonry and towers. Despite these reinforcements, Roman artillery left numerous ballistic imprints on the walls, providing modern researchers with valuable data to analyze the power and precision of Roman siege weapons [11]. Documentary evidence indicates that northern fort walls exhibit numerous cavities, which are interpreted as the result of repeated projectile impacts. These features attest to the considerable force of Roman artillery, which pulverized the wall surfaces upon impact [12,13]. Further confirmation is provided by the works of Onorato [14], who attribute these craters unambiguously to the siege conducted by Sulla in 89 BC, during which Roman ballistae (stone-throwers) and catapults (scorpiones) were deployed. Recent research on projectile impacts on the city walls during this siege highlighted the durability of Pompeii's ancient fort walls and enhanced the understanding of ancient warfare and structural resilience [11].

The work of 3D digital modelling of crater morphology carried out by Bertacchi et al. has provided valuable insight into the intended use of stone missiles and metal-tipped bolts. The analysis suggests that these weapons were designed either to compromise defensive architecture or to incapacitate defenders [15]. The diameters of the analyzed ballista projectiles ranged from 10 to 23 cm [16]. A recent analysis by Rossi supported similar conclusions, proposing that the craters observed on the walls may represent missed targets, with the intended aim being mobile shielding devices used by defenders. Her survey of the fortifications at Pompeii, constructed from Nocera Tuff, recorded penetration depths reaching 120 mm. She further described the crater morphology as cylindrical–spherical in form, with an average diameter of 140 mm and an orthogonal trajectory upon impact [11].

This study is part of an initiative to analyze the cavities identified in these fortifications as the remaining traces of the ballistic impacts from Roman artillery for their reconstruction. The study uses 3D scans of fortification cavities, collected projectiles, and material reports to develop a finite element (FE) model of the impact phenomena, inferring variables such as the impact velocity, angle, and artillery distance. The aim is to utilize the calibrated finite element (FE) model to simulate the impact outcomes for a variety of potential projectile sizes and shapes. This approach allows us to redefine the characteristics and key parameters of siege machines and artillery. By simulating different scenarios, we can gain insight into how variations in projectile dimensions and forms affect the performance and effectiveness of these historical military devices, ultimately enhancing our understanding of their capabilities and limitations.

2. Literature Review

The Sulla's siege of Pompeii in 89 BC is a pivotal event in the history of the Roman Republic, providing insights into ancient siege warfare. The northern walls of city's fortifications were especially subjected to intense bombardment from Roman artillery, including ballistae and catapults. The exceptional preservation of Pompeii, buried by the eruption of Mount Vesuvius in 79 AD, offers a unique opportunity to study the effects of these siege tactics [11].

Archaeological studies have documented the fortifications of Pompeii, revealing significant evidence of ballistic impacts from the Roman siege. Amedeo Maiuri's excavations in the early 20th century uncovered sections of the city walls, providing valuable insights into their construction and the damage sustained during the siege [17]. The northern walls exhibit numerous craters attributed to the impact of stone projectiles launched by Roman artillery. These findings are supported by Flavio Russo's experimental studies, which involved creating functioning prototypes of Roman ballistae and scorpiones, offering both theoretical and practical insights into their operation [18].

Previous research on Pompeii's defensive structures has focused on the transformation phases of the city walls, construction techniques, and the properties of materials used. Scholars such as Anniboletti [19] and Fabbri [20] have analyzed the architectural phases of the fortifications, from the archaic period to the late Samnite phase. The preservation of the walls under volcanic ash has allowed researchers to document the effects of Roman artillery, providing a basis for understanding the power and range of these ancient weapons. Advanced surveying techniques, such as photogrammetry and laser scanning, have furthered the documentation and analysis of these ballistic impacts [15].

Welded Grey Campania Tuff is a type of volcanic rock prevalent in the Campania region of Italy, particularly around the area of Pompeii. It is a form of tuff, which is a type of igneous rock formed from the consolidation of volcanic ash ejected from explosions during volcanic eruptions. The term "welded" refers to the compaction and cementation of the ash particles, which occurs due to the high temperatures and pressures associated with volcanic activity, resulting in a more dense and cohesive rock compared to other types of tuff. This material is characterized by its grey coloration and is often utilized in construction due to its relative abundance and workability [18].

The mechanical properties of Welded Grey Campania Tuff are crucial for understanding its behavior under stress, particularly in the context of historical structures like the walls of Pompeii. The density of this tuff typically ranges from 1500 to 2000 kg/m^3, with a compressive strength between 5 and 30 MP, and a tensile strength of 1 to 5 MPa. These properties are influenced by the rock's porosity, which can be as high as 30% to 50%, affecting its overall strength and durability. Studies have shown that both uniaxial compressive strength and pore collapse pressure of non-welded tuff decrease with increasing porosity [21]. This relationship is particularly important for understanding the behavior of tuff under different loading conditions. Young's modulus, a measure of the material's stiffness, generally falls between 1 and 5 GPa, while Poisson's ratio, which describes the material's deformation behavior, is usually between 0.15 and 0.30.

In the context of FEM analysis, a more accurate understanding of the mechanical properties of Welded Grey Campania Tuff is vital for accurately simulating the effects of projectile impacts on ancient fortifications. The high porosity and variable mechanical properties of the tuff necessitate detailed material characterization to ensure reliable simulation outcomes. By incorporating these properties into FEM models, researchers can better predict how such structures would have responded to the forces exerted by Roman artillery, offering valuable insights into both historical engineering practices and the effectiveness of ancient military technologies. This approach not only enhances our understanding of historical events but also contributes to the preservation and study of cultural heritage sites [15].

3. Materials and Method

The purpose of this study is to apply Finite Element Method (FEM) simulations to analyze the ballistic impacts on Pompeii's fortifications. By simulating the effects of Roman artillery, the study can provide insights into the dynamics of siege warfare, the structural integrity of ancient defenses, and the effectiveness of Roman military technology. The results provide further clarification on the relationship between the sporadic measurements and characterizations of the materials composing Pompeii's walls and their overall behavior and response to artillery ballistic impacts. The objective is to merge historical analysis with modern computational tools to enhance our insight into this historical event.

At the first step, the relevant collected mechanical properties of the Grey Tuff are reported and used to calculate missing parameters necessary for the finite element modelling of the impact.

Next, the details of the developed model, including the projectile and wall's section geometries, impact velocity, boundary conditions, and employed material models are described. The model's key aspects, such as dimensions and material properties, are based on a representative crater inside the wall that was previously analyzed and measured.

Next, the developed model is used to simulate a systematically designed series of cases to investigate the effect of impact angle on the geometry of generated indentation. The results are aimed to be implemented for further characterization of the Roman artillery, specifically the ballista.

3.1. Material Characterization

Mechanical properties of Welded Grey Campania Tuff, such as elastic modulus and unconfined compressive strength (UCS), have been extensively studied. UCS values range from 1.07 MPa to 11.68 MPa, averaging 5.23 MPa [6]. Samples from Campi Flegrei caldera showed UCS values of 10.59 MPa (dry) and 9.94 MPa (wet) [22]. Another study noted compressive stress of 4.47 MPa, decreasing to 2.59 MPa and 2.53 MPa with temperature. The tangent modulus E_t (up to 50% failure stress) is 2747 MPa, while the average and secant moduli (E_m and E_S) are 3495 MPa and 2790 MPa, respectively. Bending tests revealed UTS/UCS and E/UCS ratios of 0.286 and 561.9, respectively, indicating flexural performance [22].

Research on Pompeii's water towers reported a density of 16 kN/m^3 and a Young's modulus between 800 MPa and 1800 MPa for Nocera Tuff [23]. A study on pyroclastic flow impacts on the city walls showed a tuff wall density of 24 kN/m^3 [24]. Grey Tuff's static Young's modulus is ~1.5 GPa with a Poisson's ratio of 0.25, while dynamic properties reveal a Young's modulus of ~2.5 GPa and a Poisson's ratio of 0.20 [25]. Simulation studies have used experimental fracture energy and axial stress values to predict Grey Tuff performance under various loads, improving simulation accuracy [26]. Numerical models for tensile and compressive stress were calibrated to reflect masonry's nonlinear behavior and used concrete properties for biaxial loading stress calibration [4].

The collected Mechanical parameters of Grey Nocera Tuff, derived from the existing literature and hereafter employed in the numerical simulation for the wall, are reported in Table 1.

Table 1. Material parameters used for modelling Grey Nocera Tuff.

Parameter	Value	
UCS	4.47	MPa
E_{cm}	3495	MPa
E_{c1}	2511.7	MPa
α_α	E_{cm}/E_{c1}	-
α_d	0.4	-
E_s	2790	MPa
UTS/UCS	0.286	-
α_t	0.312·UTS	MPa

3.2. Material Modeling

In the study of tuff masonry, particularly in the Campania region of Italy, the Finite Element Method (FEM) has proven to be an invaluable tool for modeling and analyzing mechanical behavior. This approach involves creating detailed models that incorporate the material properties of tuff, such as compressive and tensile strengths, elastic modulus, and Poisson's ratio [4]. The equations reported by Nastri et. al. [4] are used for modeling tuff behavior.

For modeling tuff behavior in compression, the equations used are as follows:

$$\text{Linear behavior}: \ \sigma_c = E_{cm} \cdot \varepsilon \text{ for } \sigma_c \leq \ 50\% \text{ of UCS}, \tag{1}$$

$$\text{Nonlinear behavior}: \ \sigma_c = UCS \cdot \left[\alpha_\alpha x + (3 - 2\alpha_\alpha)x^2 + (\alpha_\alpha - 2)x^3\right] \text{ for } x \leq 1, \tag{2}$$

$$\sigma_c = UCS \cdot \frac{x}{\alpha_d \cdot (x-1)^2 + x} \text{ for } x > 1, \tag{3}$$

where $x = \frac{\varepsilon}{\varepsilon_{c1}}; \alpha_\alpha = \frac{E_{cm}}{E_{c1}}; 0.4 \leq \alpha_d \leq 4; x_{max} = 10$.

For modeling tuff behavior in tension, the equations used are as follows:

$$\text{Linear behavior}: \ \sigma_t = E_{cm} \cdot \varepsilon \text{ for } \sigma_t \leq \ \text{UTS}, \tag{4}$$

$$\text{Nonlinear behavior}: \ \sigma_t = UTS \cdot \frac{x}{\alpha_t \cdot (x-1)^{1.7} + x}, \tag{5}$$

where $x = \frac{\varepsilon}{\varepsilon_{c1}}; x_{max} = 5$.

The Concrete Damaged Plasticity (CDP) model is especially well-suited for simulating the failure mechanisms of masonry structures. It accounts for two primary failure modes: tensile cracking and compressive crushing. The evolution of damage in the material is governed by plastic deformation parameters, which are calibrated based on experimental data [4]. Due to the lack of experimental data, the CDP input parameters defining the biaxial failure surface are adopted from the literature focused on concrete modelling. The five parameters—dilatation angle (Ψ), eccentricity (ε), σ_{b0}/σ_{c0} ratio, K, and viscosity (μ)—are reported in Table 2.

Table 2. Plastic flow and yield function parameters.

Ψ	ε	K	σ_{b0}/σ_{c0}	μ
20	0.1	0.5	1.16	0

Additionally, damage evolution that affects inelastic, plastic, and crack strain are defined using the equations below, with traction and compression damage values derived from the adopted stress–strain relations. The material definition also included weight factors $w_t = 0$ and $w_c = 1$ to represent the extent of stiffness recovery during tension–compression and compression–tension switching. These parameters influence the unloading and reloading behavior of the material under complex loading conditions. For modeling damage behavior during compression (Equations (6)–(8)) and tension (Equations (9)–(11)), the equations used are as follows:

$$d_c = 1 - \frac{\sigma_i}{\sigma_{cu}}, \quad (6)$$

$$\varepsilon_c^{in} = \varepsilon_c - \varepsilon_c^{el}, \text{ where } \varepsilon_c^{el} = \frac{\sigma_c}{E_Y}; \quad (7)$$

$$\varepsilon_c^{pl} = \varepsilon_c^{in} - \frac{d_c}{(1-d_c)}\frac{\sigma_c}{E_Y}, \quad (8)$$

$$d_t = 1 - \frac{\sigma_i}{\sigma_{tu}}, \quad (9)$$

$$\varepsilon_t^{cr} = \varepsilon_t - \varepsilon_t^{el}, \text{ where } \varepsilon_t^{el} = \frac{\sigma_t}{E_Y}; \quad (10)$$

$$\varepsilon_t^{pl} = \varepsilon_t^{cr} - \frac{d_t}{(1-d_t)}\frac{\sigma_t}{E_Y}, \quad (11)$$

The CDP model's adaptability allows it to accurately represent the nonlinear behavior of masonry under both compressive and tensile loads, as demonstrated in Figures 1a and 1b, respectively. For the analysis of wall panels, the CDP model is particularly effective in controlling crack propagation by linking the evolution of the failure domain to the level of damage in the material. Nastri et. al. has employed and validated this approach through comparisons with experimental data, confirming its reliability for predicting the behavior of tuff masonry structures [4]. The methodology in this study is adapted from the study by Nastri et al. [4].

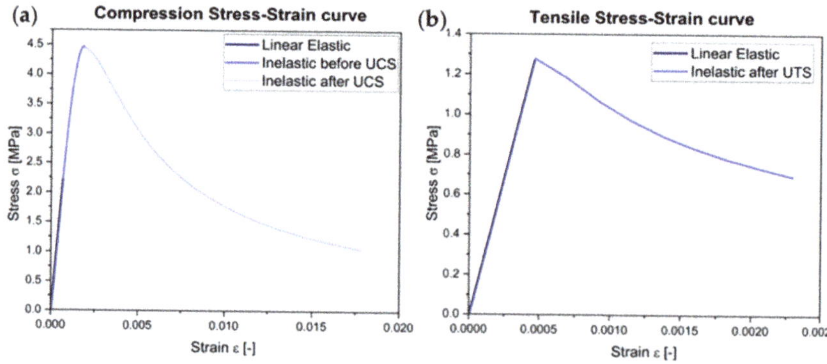

Figure 1. Stress vs. strain curve modelled using different parameters: (**a**) compression behavior and (**b**) tensile behavior.

Implementation of this model requires separate definitions of materials behaviors when put under compression and tension.

A maximum value of $x_{max} = 10$ has been selected to represent total material degradation. The parameters E_{cm} and E_{c1} are employed to characterize material response until ultimate compressive strength is reached, including initial linear elastic behavior for stresses up to 0.5 UCS and nonlinear stress–strain behavior up to UCS. Furthermore, the secant modulus E_S is utilized to approximate elastic behavior up to the material's yield strength.

For tensile stresses, $x_{max} = 5$ was utilized, accurately depicting the brittle failure characteristics of tuff under tensile loading conditions. For tensile stresses, only linear elastic behavior is considered up to the UTS point, beyond which the material rapidly fractures.

Explicit methods are often preferred for dynamic analyses or when dealing with nonlinear material behavior, as they allow for efficient computation of large models with relatively short dynamic response times. In this case, the explicit method is deemed appropriate as the simulation involves high-velocity ballistic impact of a projectile to the wall. The other general material properties of the tuff are reported in Table 3.

Table 3. Key parameters of tuff for the numerical simulation.

Parameter	Value	
Density	2297	kg/m^3
UCS	4.47	MPa
E_s	2.79	GPa
Poisson's ratio	0.2	-

The projectile was assumed to be a sphere made of basalt with only elastic behavior. Basalt's higher density, strength, and durability would make it more resistant to impact compared to the softer and more porous tuff and will show insignificant deformation [27]. Therefore, in this research, the damage to the Basalt projectile was ignored and it was modeled as a purely elastic material, with the general properties reported in Table 4.

Table 4. Key parameters of Basalt projectile for the numerical simulation.

Parameter	Value	
Density	2520	kg/m^3
UCS	60	MPa
E_s	60	GPa
Poisson's ratio	0.29	-

The interaction between different bodies was defined considering a coefficient of friction equal to 0.4 for tangential interactions.

3.3. *Numerical Analysis*

3.3.1. Impact Velocity Simulation: Ball

The simulation workflow was divided into two stages based on model dimensionality. The mesh convergence and velocity–penetration depth analyses were performed using the 2D axisymmetric model, which allowed faster simulation time while maintaining geometric accuracy for direct impacts. However, for angular impact simulations, a full 3D model was used to correctly capture asymmetries in the impact response at oblique angles and to obtain accurate width and depth values of the resulting craters.

The simulation models were developed in Abaqus FEM software [28] to simulate the impact of a basalt projectile to the tuff wall. The 3D model consists of a fully spherical projectile, with the diameter of 140 mm and a rectangular parallelepiped of 3D × 2D × 1.5D for length, width, and height, respectively, where D is the diameter of the projectile. The projectile size was adapted from a crater identified and measured from one of the studied sections of the wall. The measurement results are presented in Figure 2a–c. The overall geometry of the model is presented in Figure 3.

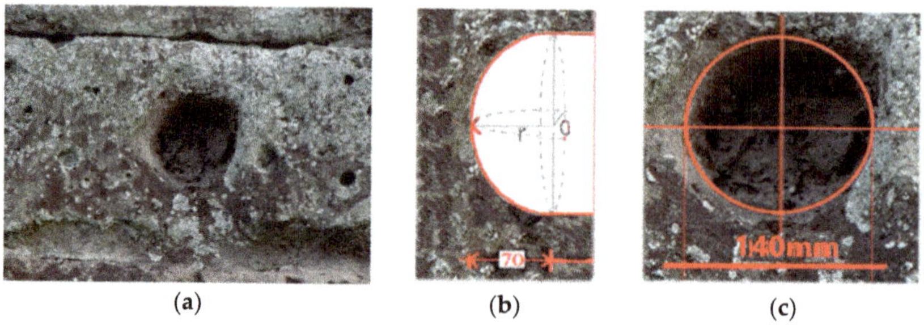

Figure 2. (**a**) Geometry of the ballistic impact crater on a block of Nocera Tuff; (**b**) side elevation resulting from (**c**) front measurements.

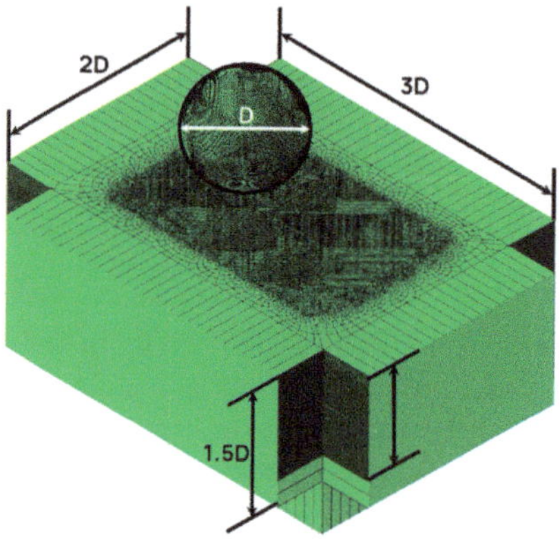

Figure 3. Geometry of the FE model (ballista).

The projectile and the wall block were discretized using a three-dimensional eight-node-brick element with reduced integration and hourglass control (C3D8R).

One row of infinite elements (CIN3D8R) was added to all the surrounding sides of the wall block. As these elements provide "quiet" boundaries to the finite element model in dynamic analyses [24], they were employed to avoid reflection of stress waves from the block's boundaries and minimize its size effect to simulate the large dimensions of the actual wall.

The developed model is then used for following analyses:

- A series of analyses with a gradually decreasing element size were carried out to analyze the mesh convergence. The projectile diameter (d) served as the reference size. A dividing factor was applied to determine the mesh size, with D/100 identified as the optimal size for damage energy convergence. Although finer mesh sizes improve stress accuracy, D/100 was deemed sufficient for estimating penetration depth.
- The model with selected appropriate mesh size was used to simulate various cases with impact velocities of 100 m/s, 80 m/s, 60 m/s, 40 m/s, 20 m/s, and 25 m/s. The penetration depth for 25 m/s closely matched the reference depth of 120 mm, with an estimated value of 121 mm. This velocity was selected in the next step.
- A series of analyses were carried out with angle of incidence gradually decreasing from 90° to 60° in 10° steps.

3.3.2. Impact Velocity Simulation: Dart

An arrowhead embedded in the fort wall provided measurable data for constructing a representative model of an arrow (dart). A CAD model of the arrowhead was prepared and used as the basis for simulation as presented in Figure 4. The corresponding crater on the wall had a penetration depth of approximately 30 mm and was considered as the reference target depth for subsequent simulation calibration. The dart is assumed to consist of an iron arrowhead and an oak-wood shaft, with a total measured length of 481 mm and total weight of 150 g. Dart parameters with their respective material densities are detailed in Table 5. Based on the tabulated density values, key physical parameters, such as the center of mass and moment of inertia, are extracted from the 3D CAD model and used for the simulation.

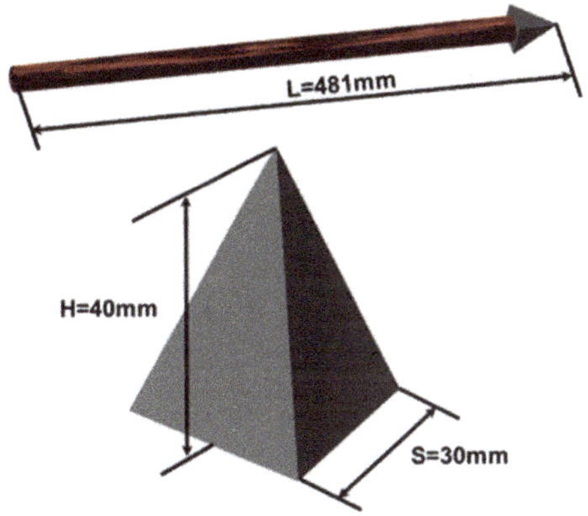

Figure 4. 3D model of the arrow(dart) with dimensions.

Table 5. Key parameters of the dart (arrow bolt).

Parameter	Value	
Length (L)	481	mm
Mass (m)	150	g
Arrowhead material density	7800	kg/m^3
Wooden shaft density	640	kg/m^3

The simulation substrate was developed with dimensions of 3S × 2S × 1.5S, where S = 30 mm represents the base length of the arrowhead. The geometric layout and key dimensions of the model are illustrated in Figure 5.

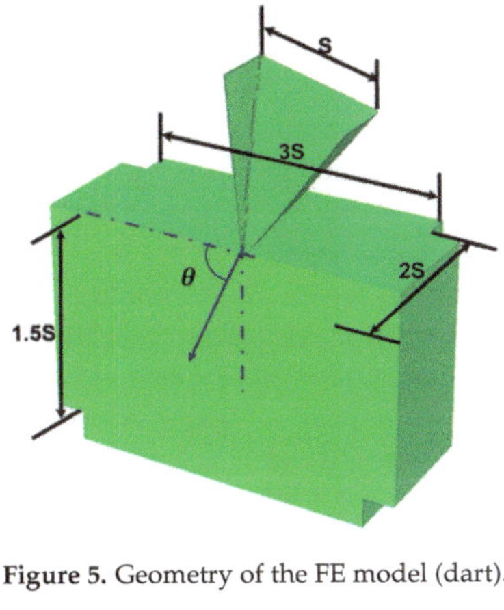

Figure 5. Geometry of the FE model (dart).

The dart and the wall block were discretized using a three-dimensional, quadrilateral (R3D4) and triangular (R3D3), and three-dimensional eight-node-brick element with reduced integration and hourglass control (C3D8R), respectively.

The tuff target domain was extended along all the non-impact faces and assigned infinite elements (CIN3D8) to replicate realistic stress wave propagation and avoid wave reflection from boundaries.

- For the material model, the arrowhead was considered significantly stiffer than the tuff and was considered as a rigid body, while the tuff wall was modelled as a deformable continuum with non-linear elastic–plastic behavior. To mitigate numerical singularities typically associated with sharp features in FEM simulations, two fillet radii, R = 5.6 mm and R = 1 mm, were evaluated with an initial impact velocity of 10 m/s. The radius was found to significantly influence penetration depth, with R = 5.6 mm and R = 1 mm resulting in a penetration depth of 28.086 mm and 43.031 mm, respectively. Due to the excessive computational cost for smaller radii, a value of R = 1 mm was fixed for subsequent simulations.
- A mesh convergence study was performed by varying mesh size factors (40, 50, 75, 100, and 125). The optimal mesh factor was determined as 50 (S/50 as the element size), providing a balance between computational cost and penetration accuracy.
- To identify the most representative velocity for reference penetration depth, simulations were conducted at 6, 8, and 10 m/s under the normal (90°) impact condition. The impact velocity of 6 m/s was found to closely match the reference penetration of 30 mm and was selected for subsequent studies.
- Further simulations were carried out to study the influence of oblique impacts using the selected velocity of 6 m/s. Impact angles of 90°, 80°, 75°, 70°, and 60° were considered to capture variations in penetration behavior and impact mechanics under changing incident directions.

4. Results

4.1. Mesh Convergence Analysis

To ensure accurate representation of projectile–tuff wall interaction, mesh convergence study was performed for both ball and dart impact simulations. The projectile's characteristic dimensions—diameter (D) for the ball and arrow tip base length (S) for the dart—were used as reference lengths to define mesh sizes.

For the different mesh size factors considered, F_S = 100 with mesh element size of D/100 was identified as the optimal resolution for ball impact model. As the mesh refinement was increased, the cavity created by the projectile became smoother and more refined, with reduced surface irregularities and element-based roughness. This improvement in the geometric clarity of the crater is evident from the mesh convergence (Figure 6a). Although further refinement offered better stress field resolution, it introduced higher computational costs without significantly improving penetration depth accuracy. Thus, D/100 was selected as an optimal element size between computational efficiency and result accuracy.

The sharp edges of the square pyramid shaped dart introduced potential numerical instabilities during mesh refinement simulations, and a more nuanced approach was required. From the Figure 6b, it is evident that on reducing element size, some stability in penetration depth was noted at S/50 and S/75. However, beyond S/75, the penetration depth began to increase inconsistently with continued refinement. This was interpreted as a possible onset of numerical instability due to the sharp geometry and the limits of contact and damage algorithms in the model. As a result, S/50 was selected for subsequent simulations, offering sufficient resolution for capturing crater morphology while avoiding instability and excessive computational cost.

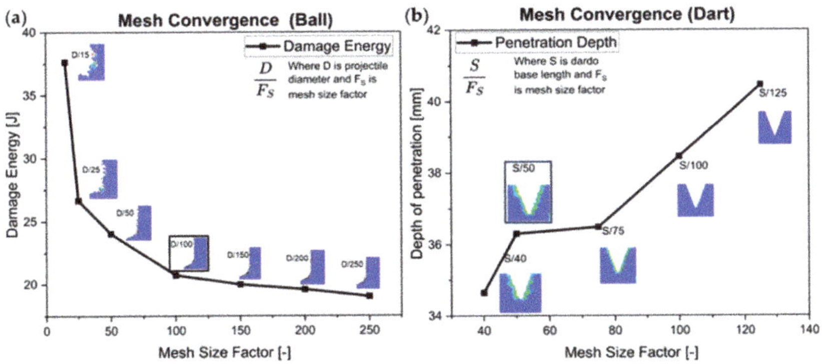

Figure 6. Mesh convergence analysis for (**a**) ball, considering damage energy to compare different mesh sizes, and (**b**) dart, considering penetration depth to compare different mesh sizes.

4.2. Velocity Range

Simulations were conducted to evaluate the relationship between impact velocity and penetration depth. The goal was to estimate the velocity at which the simulation results align with archaeological observations of crater depths of approximately 120 mm for the ball and 30 mm for the dart.

A series of simulations were conducted for the ball–tuff interaction at impact velocities of 20, 25, 40, 60, 80, and 100 m/s. The simulated crater depth values as a result of varying velocities are reported in Table 6 and resulting trend of the penetration depth curve is presented in Figure 7a following a near-parabolic curve. This behavior highlights increasingly higher velocity increments are required to achieve proportional gains in depth. Among the tested cases, penetration depth from velocity of 25 m/s closely matched with the reference archaeological value of 120 mm.

Table 6. Simulation results for the penetration depth at different impact velocities of the ball projectile.

Velocity [m/s]	20	25	40	60	80	100
Penetration Depth [mm]	100.4	121	151.7	184.2	213.3	228.5

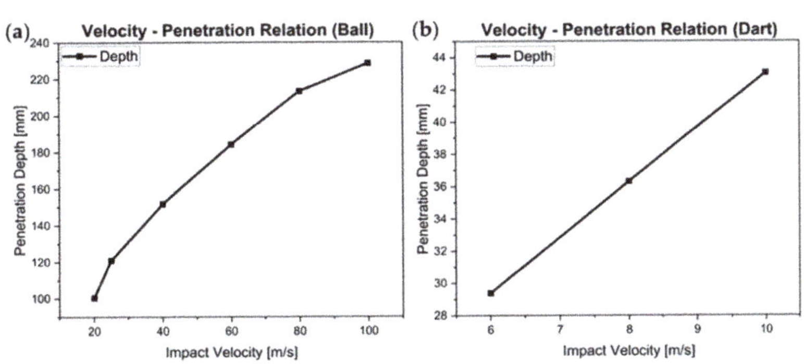

Figure 7. Penetration depth vs. impact velocity curve for (**a**) the ball projectile and (**b**) the dart.

Similarly, velocities of 6, 8, and 10 m/s were simulated for the dart–substrate impact case, final penetration depth values are reported in Table 7 and the resulting trend is presented in Figure 7b. The penetration depth from a 6 m/s velocity was found to be approximately 30 mm, matching the observed crater depth in Pompeii's northern fort walls. This value was subsequently adopted as the calibrated impact velocity for the arrow.

Table 7. Simulation results for the penetration depth at different impact velocities of the dart.

Velocity [m/s]	20	25	40
Penetration Depth [mm]	100.4	121	151.7

4.3. Angular Impact Simulation

A 3D simulation model was created to study Grey Tuff's response under angled impacts, with calibrated velocities of 25 m/s for the ball and 6 m/s for the dart. Impact angles of 90°, 80°, 70°, 60°, and 75° were considered with 90° indicating a direct hit. Crater dimensions were quantified by measuring displacement of the center of mass of the projectile in both perpendicular and lateral directions from the impact surface as reported in Tables 8 and 9 for ball projectile and dart, respectively. Figure 8a,b present crater width and depth curve with respect to different angles, showing material response is similar in both the cases. Key findings include the following:

- The crater's width and depth showed a direct relationship with the respective velocity components, where the root of the quadratic sum of the two components corresponded to the penetration depth observed under perpendicular impact conditions.
- This correlation demonstrated the material's uniform reaction to impacts from various angles.

Table 8. Crater width and depth for different impact angles of the ball projectile.

Impact Angle [°]	**Width [mm]**	**Depth [mm]**
60	44.3522	85.9216
70	28.1578	92.0044
75	20.8203	94.0796
80	13.7902	95.3911
90	0	96.2131

Table 9. Crater width and depth for different impact angles of the dart.

Impact Angle [°]	**Width [mm]**	**Depth [mm]**
60	18.4178	25.6678
70	11.0858	26.3882
75	8.7181	27.6828
80	5.0315	28.1649
90	0	29.3830

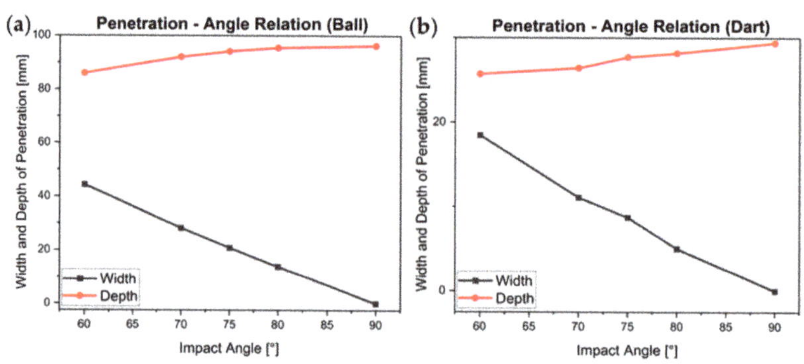

Figure 8. Crater width and depth vs. impact angle for (**a**) the ball projectile and (**b**) the dart.

By combining experimental data, numerical modeling, and simulation outcomes, this approach effectively characterized the impact resistance and failure mechanisms of Grey Campania Tuff under ballistic conditions.

Depth of Penetration Assessment: For the ball, an axisymmetric model initially determined an optimal velocity of 25 m/s, producing a penetration depth of approximately 121 mm, closely aligning with the archaeological benchmark of 120 mm. When simulated using the 3D model, however, the penetration depth was slightly lower, which may be attributed to energy dissipation across additional degrees of freedom and boundary effects within the full 3D domain.

Evaluation of Angled Impacts: The simulation outcomes confirmed that crater depth and width closely followed the decomposition of velocity into normal and tangential components. The square root of the sum of their squares consistently aligned with the total penetration depth for perpendicular impact. This observation validates the assumption that Grey Campania Tuff exhibits a uniform directional response under varying impact angles. Although slightly higher variation was observed in the case of dart model results across different angles. This can be attributed to the pyramid geometry of the arrowhead causing different net volume engagement with the substrate at oblique angles, thereby affecting the depth of penetration.

Simulating Material Response: The numerical approach effectively captures the nonlinear compression and tension characteristics of Grey Campania Tuff, offering dependable parameters for subsequent simulations.

5. Discussion and Conclusions

The simulation results offer valuable insight into Pompeii fort wall resilience to the stones and arrows projected by the ancient roman artilleries. First, the mesh convergence study showed that using a mesh size of D/100 for the ball gave smooth and realistic craters without needing too much computing power. While, the sharp edges of the dart can cause numerical instabilities, S/50 was chosen as a stable and computationally efficient mesh size.

The velocity–penetration relationship showed a nonlinear, near-parabolic trend for both projectile types. A ball projectile with impact velocity of 25 m/s had a penetration of 121 mm, closely matching the archaeological benchmark of 120 mm, while a dart velocity of 6 m/s produced depth of approximately 30 mm, which is consistent with observed wall damage. These values were used as a reference for the angular impact simulations.

Angular impact analyses revealed that the crater width increased with decreasing angle and penetration depth reduced proportionally but the overall displacement remained nearly constant for the ball. This indicated uniform energy dissipation across angles. The dart displayed slightly more variation in total displacement due to its pyramid shape, altering the penetrated volume at oblique angles. Additionally, 3D simulations of the ball projectile showed slightly reduced penetration depths compared to the 2D axisymmetric model, likely due to added lateral confinement and redistribution of stress not captured in simpler representations.

The research produced a robust and versatile 3D finite element (FE) model accurately depicting material properties for simulating wall reactions to siege engine projectiles. This model was validated with documented experimental data and observations. The FE model can analyze wall responses to impacts from projectiles of different shapes and materials and contribute to a better understanding of the design of the Roman war machines.

Author Contributions: M.M.T.: conceptualization, finite element analyses, data curation, and writing—review and editing. A.A.L.: conceptualization, supervision, and writing—review and editing. M.G.: writing—review and editing, supervision, resources, and funding acquisition. All authors have read and agreed to the published version of the manuscript.

Funding: This research was funded by the European Union—NextGenerationEU, M4C2 I1.1, Progetto PRIN 2022 "SCORPiò-NIDI", Prot. 20222RJE32, CUP. The authors would like to express their gratitude to the management and the appointed officials of the offices of the Pompeii Archaeological Park for granting authorizations for site access and survey operations.

Institutional Review Board Statement: Not applicable.

Informed Consent Statement: Not applicable.

Data Availability Statement: Research data are available upon request.

Acknowledgments: The author would like to thank Gabriel Zuchtriegel, Director of the Archaeological Area of Pompeii, Giuseppe Scarpati, Head of the Study and Research Area, and Valeria Amoretti. The research activities are part of the MUR—PRIN 2022 project "SCORPiò-NIDI".

Conflicts of Interest: The authors declare no conflicts of interest. The funders had no role in the design of the study; in the collection, analyses, or interpretation of data; in the writing of the manuscript; or in the decision to publish the results.

References

1. Dobbins, J.J. Problems of Chronology, Decoration, and Urban Design in the Forum at Pompeii. *Am. J. Archaeol.* **1994**, *98*, 629–694. [CrossRef]
2. McNicoll, A.W.; Milner, N.P. *Hellenistic Fortifications from the Aegean to the Euphrates*; Oxford University Press: Oxford, UK; New York, NY, USA, 1997.
3. Scherer, A.K.; Golden, C. Tecolote, Guatemala: Archaeological Evidence for a Fortified Late Classic Maya Political Border. *J Field Archaeol.* **2009**, *34*, 285–305. [CrossRef]
4. Nastri, E.; Tenore, M.; Todisco, P. Calibration of concrete damaged plasticity materials parameters for tuff masonry types of the Campania area. *Eng. Struct.* **2023**, *283*, 115927. [CrossRef]
5. Langella, A.; Bish, D.; Calcaterra, D.; Cappelletti, P.; Cerri, G.; Colella, A.; Gennaro, R.; Graziano, S.; Perrotta, A.; Scarpati, C.; et al. L'Ignimbrite Campana (IC). In *Le Pietre Storiche della Campania*; de Gennaro, M., Calcaterra, D., Langella, A., Eds.; Luciano Editore: Napoli, Italy, 2013; pp. 155–178.
6. Piovesan, R.; Maritan, L.; Meneghin, G.; Previato, C.; Baklouti, S.; Sassi, R.; Mazzoli, C. Stones of the façade of the Sarno Baths, Pompeii: A mindful construction choice. *J. Cult. Herit.* **2019**, *40*, 255–264. [CrossRef]
7. Jackson, M.; Deocampo, D.; Marra, F.; Scheetz, B. Mid-Pleistocene pozzolanic volcanic ash in ancient Roman concretes. *Geoarchaeology* **2010**, *25*, 36–74. [CrossRef]
8. Jackson, M.D.; Landis, E.N.; Brune, P.F.; Vitti, M.; Chen, H.; Li, Q.; Kunz, M.; Wenk, H.R.; Monteiro, P.J.M.; Ingraffea, A.R. Mechanical resilience and cementitious processes in Imperial Roman architectural mortar. *Proc. Natl. Acad. Sci. USA* **2014**, *111*, 18484–18489. [CrossRef] [PubMed]
9. De Luca, R.; Miriello, D.; Pecci, A.; Domínguez-Bella, S.; Bernal-Casasola, D.; Cottica, D.; Bloise, A.; Crisci, G.M. Archaeometric Study of Mortars from the Garum Shop at Pompeii, Campania, Italy. *Geoarchaeology* **2015**, *30*, 330–351. [CrossRef]
10. Heap, M.J.; Farquharson, J.I.; Kushnir, A.R.L.; Lavallée, Y.; Baud, P.; Gilg, H.A.; Reuschlé, T. The influence of water on the strength of Neapolitan Yellow Tuff, the most widely used building stone in Naples (Italy). *Bull. Volcanol.* **2018**, *80*, 51. [CrossRef]
11. Rossi, A. The Survey of the Ballistic Imprints for a Renewed Image of Unearthed Pompeii. *Nexus Netw. J.* **2024**, *26*, 307–324. [CrossRef]
12. Van Buren, A.W. Further Studies in Pompeian Archaeology. *Mem. Am. Acad. Rome* **1925**, *5*, 103–113. [CrossRef]
13. Van Buren, A.W. Further Pompeian Studies. *Mem. Am. Acad. Rome* **1932**, *10*, 7–54. [CrossRef]
14. Maiuri, A. Introduzione allo studio di Pompei. In *Corso di Antichità Pompeiane ed Ercolanesi 1942-43*; Onorato, G.O., Ed.; G.U.F. "Mussolini": Napoli, Italy, 1943.
15. Bertacchi, S.; Barsanti, S.G.; Rossi, A. Geometry of Wall Degradation: Measuring and Visualising Impact Craters in the Northern Walls of Pompeii. *SCIRES-IT—SCIentific Res. Inf. Technol.* **2024**, *14*, 111–128. [CrossRef]
16. Piranesi, F.; Piranesi, G.B.; Guattani, G.A. Pompeia Antiquities. In *Antiquity of the Grande Grece, from the Royaume de Naples*; Beaux-arts de Paris: Paris, France, 1804; pp. XXXVI–LXXII.
17. Maiuri, A. *Studi e ricerche sulla fortificazione di Pompei*; G. Bardi: Milano, Italy, 1929.
18. Russo, F.; Russo, F. *89 a.C.: Assedio a Pompei: La dinamica e le tecnologie belliche della conquista sillana di Pompei*; Flavius Edizioni: Napoli, Italy, 2005.
19. Anniboletti, L. Le fasi delle fortificazioni di Pompei. *SIRIS Studi Ric. Scuola Spec. Beni Archeol. Matera* **2015**, *15*, 49–70.
20. Fabbri, M. The city-wall in the Roman Age. *Constr. Hist.* **2021**, *36*, 1–20.

21. Zhu, W.; Baud, P.; Vinciguerra, S.; Wong, T. Micromechanics of brittle faulting and cataclastic flow in Alban Hills tuff. *J. Geophys. Res. Solid Earth* **2011**, *116*, B06209. [CrossRef]
22. Manfredi, G.; Marcari, G.; Voto, S. Analisi e caratterizzazione meccanica di murature di tufo. In Proceedings of the 15th CTE Congress, Bari, Italy, 4–6 November 2004; pp. 4–6.
23. Lorenzoni, F.; Valluzzi, M.R.; Salvalaggio, M.; Minello, A.; Modena, C. Operational modal analysis for the characterization of ancient water towers in Pompeii. *Procedia Eng.* **2017**, *199*, 3374–3379. [CrossRef]
24. Ruggieri, N.; Galassi, S.; Tempesta, G. The effect of pyroclastic flows of the 79 AD eruption of Mount Vesuvius on the Pompeii's city walls. The case study of the sector near the Tower XI. *J. Cult. Herit.* **2020**, *43*, 235–241. [CrossRef]
25. Heap, M.J.; Baud, P.; Meredith, P.G.; Vinciguerra, S.; Reuschlé, T. The permeability and elastic moduli of tuff from Campi Flegrei, Italy: Implications for ground deformation modelling. *Solid Earth* **2014**, *5*, 25–44. [CrossRef]
26. Salvalaggio, M.; Roca Fabregat, P.; Valluzzi, M.R.; Lorenzoni, F. Finite element micro-modelling for the characterization of inclined head joints archaeological masonry: The case of Villa Diomede in Pompeii. In Proceedings of the COMPDYN 2017—6th International Conference on Computational Methods in Structural Dynamics and Earthquake Engineering, Rhodes Island, Greece, 15–17 June 2017; National Technical University of Athens: Rhodes Island, Greece, 2017; pp. 2460–2469.
27. Yan, F.; Feng, X.T.; Chen, R.; Xia, K.; Jin, C. Dynamic Tensile Failure of the Rock Interface Between Tuff and Basalt. *Rock Mech. Rock Eng.* **2012**, *45*, 341–348. [CrossRef]
28. A.U. Manual. *Abaqus User Manual*; Abacus: New York, NY, USA, 2020.

engineering proceedings

MDPI

Proceeding Paper

Ancient Projectile Identification Through Inverse Analysis: Case Studies from Pompeii †

Simone Palladino, Renato Zona and Vincenzo Minutolo *

Department of Engineering, University of Campania "L. Vanvitelli", 81031 Aversa, Italy; simone.palladino@unicampania.it (S.P.); renato.zona@unicampania.it (R.Z.)
* Correspondence: vincenzo.minutolo@unicampania.it

† Presented at the Conference "Discovering Pompeii: From Effects to Causes—From Surveying to the Reconstructions of Ballistae and Scorpiones", Aversa, Italy, 27 February 2025.

Abstract: A straightforward method for determining the causes of impact relics left by ancient projectiles on the city walls of Pompeii is proposed based on principles of plasticity and fracture mechanics. The inverse analysis begins with the measured craters caused by spherical projectiles or darts launched by the Roman army during the siege of 89 B.C. A Mathematica© notebook is presented, enabling the calculation of projectile impact velocity from the known dimensions of the projectiles and the mechanical properties of the wall material.

Keywords: inverse analysis; ancient ballista; fracture release rate; plastic dissipation

1. Introduction

The work is aimed at evaluating the characteristics of ancient war machines such as the ballista in Roman times. The study is focused on the relic of the impact of projectiles on some of Pompei's walls that were the object of siege during Silla's campaign in 89 B.C. [1,2]. The geometry of impact craters offers essential clues about the dynamics of the impacting projectiles. Inverse analysis is implemented to relate the estimated energy dissipation for producing impact craters and the momentum of the projectiles. In the outer walls of the City of Pompei, two principal signs of impact can be recognized: the first, a large crater of about 140 mm, seems to be created by the impact of a spherical projectile from a fixed ballista; the second, a small conical hole, with the diameter of about few centimeters, $\phi_d = 30$ mm, seems to be caused by an ancient automatic portable ballista able to shot light darts whose mass was $m_d = 150$ g called *polybolos* [3–8]. The main parameters one should define to carry out the analyses are the mechanical properties of the wall of the City. The material constituting the structure is grey tuff, of the Neapolitan area, namely Nocera's tuff. Mechanical properties of the material suggest addressing the derivation of the breakage parameters of the material from analogous studies made on the resistance and the collapse of soils and rocks [9–11]. Work [12] employs the Mohr–Coulomb failure criterion and limit analysis via a paraboloidal collapse surface to derive closed-form solutions for masonry structure collapse and energy dissipation. Similarly, refs. [13,14] emphasize how porosity and heterogeneity affect collapse loads and apply a limit analysis procedure for evaluating the collapse load of masonry domes. Recent studies have also focused on the fracture behavior of volcanic rocks as natural materials and as reinforcing phases in composites. Fracture toughness, often quantified via the critical energy release rate (G_f) or the stress intensity factor (Sif), serves as a fundamental indicator of a material's resistance to crack propagation under linear elastic fracture mechanics. The propagation and coalescence of fractures are studied in several papers [15–19] where the propagation and the load-bearing capacities of structures made of heterogeneous and no tension materials are presented.

By analyzing post-impact deformation and fracture patterns, in the paper, the velocity of projectiles quantifying momentum transfer and impact energy absorption is analyzed. Special emphasis is placed on evaluating energy dissipation via localized cracking and plastic energy accumulation at fracture sites. Ballista spherical projectiles and dart projectiles of portable ballista effects are simulated using different strategies that perform the calculation based on a static approach that results faster and more robust with respect to transient dynamics numerical analysis. Through the described methods, we calculate the projectile velocity and the energy dissipation produced by shooting from the measured impact holes directly.

The paper presents the results of a finite element program under static loading, simulating the applied force due to the impact on a point of the wall. The material of the wall is grey tuff, and the projectile is described with the equivalent one-point force. The intensity of the force at the collapse of the application point neighborhood is evaluated, and the permanent strain amount is used to calculate dissipated energy corresponding to the transformation of the kinetic energy of the projectile. In a second approach, the fracture energy required for breaking the material volume of the wall at the impact target, whose dimensions were detected in [3–5], is related to kinetic energy as well to obtain projectile impact velocity.

Finally, the elastic energy accumulated during the loading process in the impact is compared with the fracture release energy to verify the balance in the framework of the linear fracture mechanics theory.

Conclusions and discussion highlight the feasibility of the strategies that are able to evaluate compatible velocity and the momentum of the projectiles of the ancient Roman war machines during I century B.C.

2. Materials and Methods

The inverse analysis of the effects of ancient projectiles on the wall structures of cities is described considering direct energy evaluation. Numerical calculations are the main strategy nowadays used due to the great affordably of FEM modeling. However, a direct calculation based on simple models that use mechanics and descriptions of the phenomena can give valuable indications on the order of magnitude and straightforward results to make an inverse analysis.

The impact of the projectile on the wall dissipates energy in several manners; the most is ascribed to the energy to produce irreversible deformation. To evaluate the dissipated energy, two strategies are implemented in the work.

Ductile Plastic Dissipation by FEM

At first, we implemented static calculation using an elastoplastic model of ductile dissipation. The wall structure was loaded with one nodal force up to the structure collapse that involved locally an imitated number of elements, as is to be expected in a point load problem. The irreversible strain at the incoming collapse measures the stress–strain dissipation, namely the deviatoric stress acting on the elements where the plastic strain accumulates so works that cannot be recovered. The resulting plastic dissipation is equated to the kinetic energy of the projectile resulting in its velocity.

The dimensions of the wall were considered from the measures reported in [4]. A preliminary study was conducted to recognize the mesh size and the wall dimensions so that boundary effects could be neglected. The geometric properties of the wall and the corresponding mechanical parameters are shown in Table 1.

Table 1. Mechanical properties of the target wall.

Young Modulus (E)	2914 MPa
Poisson Coefficient (ν)	0.29
Yield Stress (σ_Y)	4.47 MPa

At the incoming collapse, the plastic dissipation can be evaluated through the permanent strain and corresponding yield stress.

$$D^p = \int_B \int_{t_0}^{t_c} \sigma_0 \dot{\varepsilon}^p dt dV = \int_B \sigma_0 \varepsilon^p dV \tag{1}$$

Equation (1) is obtained assuming the normality flaw rule and perfect plasticity; hence, the deviatoric stress is constant in time, and permanent strain reduces to accumulated permanent strain that has no components along with the hydrostatic stress line.

In the FEM result, the permanent strain deviatoric norm was multiplied for the deviatoric stress yielding and integrated over the element volume, resulting in

$$D^p = \sum_{e=1}^{pe} \sigma_{de} \varepsilon_{de}^p V_e \tag{2}$$

By considering the complete transformation of the kinetic energy of the dart into plastic dissipation, one can evaluate

$$\frac{1}{2} m_d v_d^2 = D^p \tag{3}$$

that results in

$$v_p = \sqrt{\frac{2D^p}{m_d}} \tag{4}$$

The second approach is based on the fracture energy capable of breaking the projectile target volume into pieces. The involved volume is derived from the dimensions of the measured holes detected in the observation of ancient relics.

The inverse analysis allowed us to recognize the velocities of the projectile from the probable dimensions of fragments.

The procedure is based on some hypotheses, namely that the fragments have a polyhedral shape; in particular, an icosahedral shape was assumed, and the surface area of the polyhedron was calculated considering the presence of perturbation of the side plane. The perturbation was modeled by using pyramidal protrusions. It is known that the pyramid side surface is $\beta = \sqrt{5} = 2.24$ times the base area. Hence, the amplification coefficient β was introduced, yielding to the evaluation of the effective surface of the crack opening to break the target into pieces

$$S_f = \beta \cdot 5\sqrt{3}\phi^2 \tag{5}$$

where ϕ is the dimension of the face of the icosahedron. The relationship that links the diameter of the sphere that is circumscribed to the icosahedron D_s is

$$f = \sqrt{0.5 - \frac{\sqrt{5}}{10} D_s} \tag{6}$$

In Equation (5), the surface refers to a single fragment, and the fragment volume is given by

$$V_f = \frac{5}{12}\left(\sqrt{5} + 3\right)\phi^3 \tag{7}$$

Equation (7) allows calculating the number of fragments in the volume of the detected hole.

$$V_H = \pi d_H^2 \left(\frac{D}{2} - \frac{d_H}{3} \right) \tag{8}$$

$$n_f = \frac{V_H}{V_f} \tag{9}$$

Finally, the dissipated energy due to crack openings is

$$E_f = n_f G_f \frac{S_f}{2} \sqrt{5} \tag{10}$$

where the denominator 2, under the surface term S_f, accounts for the presence of two surfaces at any crack openings.

To obtain the sought impact velocity, the energy (10) is equated to the kinetic energy of the projectile

$$k = \frac{1}{2} m_p v_p^2 \tag{11}$$

so that

$$v_p = \sqrt{\sqrt{5} \frac{n_f G_f S_f}{m_p}} \tag{12}$$

3. Results

The procedures described in the previous section were applied to the structure of Pompei's wall reported in [4]. The materials' properties and the geometry were derived from the literature, and the two approaches that can be summarized as ductile and fragile are reported hereafter.

3.1. FEM Ductile Plastic Analysis

The FEM model simulates the target wall whose properties are reported in Table 1.

The FEM analysis was carried out to obtain the final configuration of the local collapse due to the impact of a dart with a mass of 0.150 kg using a step-by-step collapse calculation. The collapse load was calculated as follows:

$$F_c = 20 \text{ kN} \tag{13}$$

The geometry, mesh, and von Mises stress are depicted in Figure 1, where the stress concentration at the incoming collapse around the impact area is highlighted.

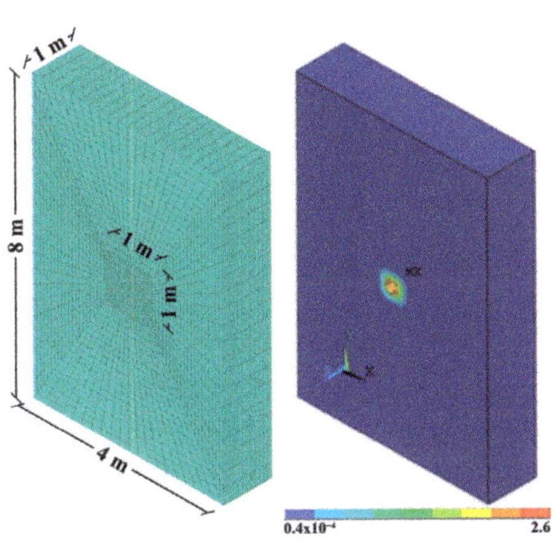

Figure 1. Finite element mesh refinement prescribed in the impact area (**left**). The contour plot of the von Mises stress, corresponding to the incoming collapse around the impact area (**right**).

The FEM calculation furnished plastic dissipation D^P at the incoming collapse given by

$$D^P = 15.91 \text{ kJ} \tag{14}$$

that allows calculating the dart velocity from Equation (4) as

$$v_D = \sqrt{\frac{2 \cdot 15.91}{0.150}} = 14.56 \text{ m s}^{-1} \tag{15}$$

3.2. *Fragile Procedure via Fracture Energy Release*

The fracture energy release of the material constituent of the wall, G_f, is obtained from literature reports about volcanic materials [13]; it is calculated from the critical Stress Intensity Factor $KI_c = 1.5 \text{ MPa} \sqrt{m}$:

$$G_f = \frac{KI_c^2}{E} = 0.81 \text{ kJ m}^{-2} \tag{16}$$

3.3. *The Ballista Spherical Projectile*

As a first application of the method, the case of a projectile of a ballista constituted by a basalt sphere was considered.

Ballista projectile impact over the wall produced a hemispherical hole whose surface is characterized by several fragment tracks. The average dimension of the fragments was detected and reported in the reconstruction made in [5].

The geometric characteristics of the impact track are reported in Table 2.

Table 2. The geometry of the projectile and relics on the wall.

Diameter D	Density ρ	Hole Depth d_H	Hole Volume V_H	Mass m_p
140 mm	2.800 kg m^3	120 mm	$1.149 \cdot 10^{-2}$ m^3	3.22 kg

The fragment radius is considered by observation of the dimensions of tracks on the 3D reconstruction of the shots imaging reported in [4]. In Figure 2, a sample of the 3D reconstruction is depicted. The yellow drawings in the picture highlight the shape and the dimensions of probable fragments of the material destroyed by the shooting. The dimensions of the pieces can be evaluated considering that the projectile marks on the wall have the diameter reported in Table 2 D = 140 mm. Although the marks have been smoothed by aging, the remaining signs allow recognizing an average dimension between 20 mm and 35 mm.

Relationship (12) has been applied to different fragments' dimensions in the range $0.1 \text{ mm} \le D_s \le 40 \text{ mm}$, yielding to the diagram reported in Figure 3.

The diagrams show that for fragment dimensions of about $20 \text{ mm} < D_s < 30 \text{ mm}$, which are the most probable to fit the measured cavity in the wall, the average projectile speed is about $15.19 \text{ ms}^{-1} < v_s < 12.40 \text{ ms}^{-1}$ that corresponds to energy

$$464.0 \text{ J} < E < 309.3 \text{ J} \tag{17}$$

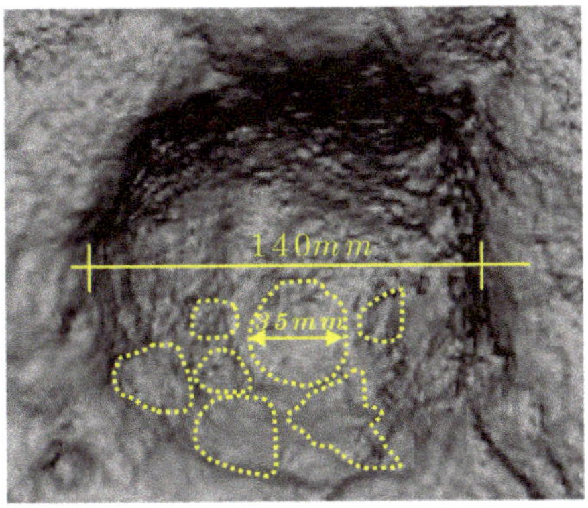

Figure 2. Three-dimensional representation of the shooting mark. Yellow indicates the traces and dimensions.

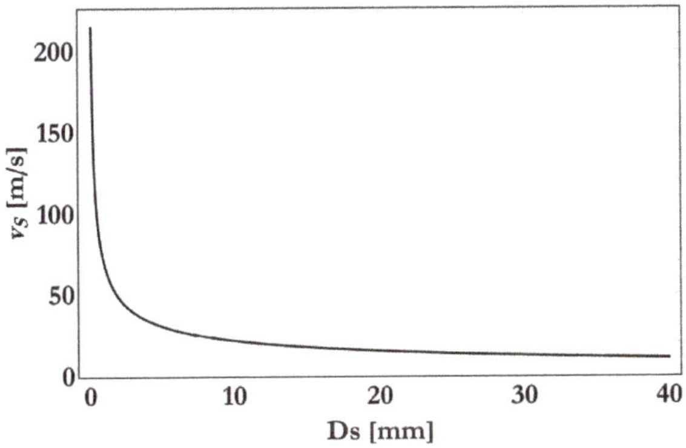

Figure 3. Spherical projectile effect: velocity versus the diameter of fragments.

3.4. Darts from Portable Ballista or Polybolos

The analysis of the effect of darts follows the same approach used for the spherical projectiles but with the modification required by the shape of the projectile tip that is pyramidal. The geometry of the dart is reported in Table 3.

Table 3. Geometry of the darts and their relics on the wall.

Base B	Height H	Density ρ	Hole Dept d_H	Hole Volume V_H	Mass m_d
30 mm	40 mm	7.800 kg m^3	40 mm	$8.559 \cdot 10^{-6}$ m^3	0.15 kg

Using the same equation as for the spherical projectile, except for the hole shape and volume, the following representation of the velocity–fragment diagram is obtained (see Figure 4).

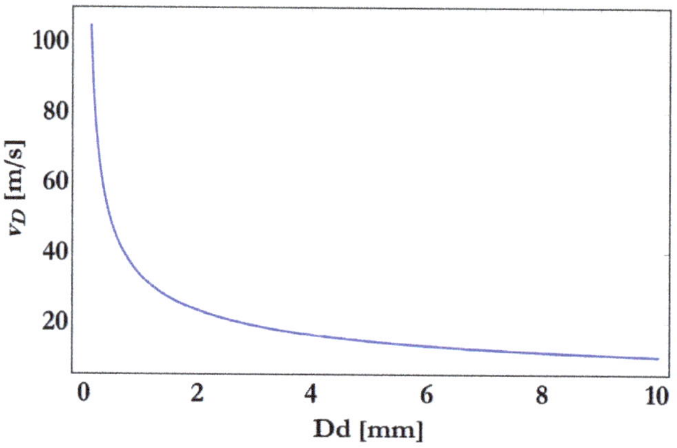

Figure 4. Dart effect: velocity of projectile versus the diameter of prevised fragments.

The results summarize the dart effect as follows:

$$4\ \text{mm} < \text{D}_\text{d} < 6\ \text{mm} \rightarrow 16.54\ \text{ms}^{-1} < \text{v}_\text{D} < 13.50\ \text{ms}^{-1} \tag{18}$$

that corresponds to the energy

$$20.52\ \text{J} < \text{E} < 13.67\ \text{J} \tag{19}$$

A final consideration about the fracture energy release method can be made considering that the energy release for the fracture to develop can be associated with strain energy recovery due to the relaxation of the stress state at the impact. The idea follows from the Griffith energy release approach to fracture. We can assume that, on impact, the projectile loaded the structure by a force consequence of the deceleration. The effect is a deformation of the wall that produces elastic energy accumulation in the structure. If the accumulated elastic energy is released through fracture openings, the dissipation equates with the elastic energy. Such an approach is not suitable for identification purposes since we have no information on the velocity and energy the projectile had at the impact stage. However, let us consider the force intensity equal to the collapse calculated by plastic collapse analysis. We can evaluate the strain energy at the incoming collapse, prior to plastic strain occurring, and consider the dissipated energy due to the fracture of the target as the energy release due to crack openings.

4. Discussion and Conclusions

The presented research study describes the inverse analysis of projectiles from the effects measured on the ancient Roman wall of the City of Pompei. The city walls, submerged by the Vesuvius' eruption in 79 AD, have been preserved, and the tracks of ballista shots of the Silla's army in 86 B.C. The study of the tracks is carried out considering two types of dissipative mechanisms, one due to ductile plastic dissipation through permanent deformation yielding to target destruction, and the other T through considering the fracture energy release for the target fragments to separate and cause the target hole we found in the city fortification.

The proposed models are compared with the numerical calculation performed using the commercial FEM program Ansys ©, where the direct step-by-step analysis is implemented till the structural collapse. The collapse obtained through finite elements plastic analysis offers the limit load corresponding to the rupture of the target in the wall. It has to be stressed that the obtained results, even if corresponding to direct calculation in static mode, highlight that the plastic deformation is localized at the impact point and does not involve a large distance. Consequently, the dissipated strain energy at the collapse is bound and can be easily calculated. The only drawback of such a calculation remains in the fact that no upper bound of the ductility of the structure is considered. A more detailed dynamic step-by-step analysis that considers the plastic strain evolution and impact dissipation together with the softening behavior of the crushing mechanism of fragile materials that employ a large amount of machine time and memory describes the mechanism with more detail. However, the approach proposed here offers acceptable results, as it can be seen by comparison with work [3].

The simplified analysis represented in Equations (4) or (12) offers acceptable results with respect to both FEM calculation and [3]. The simplified procedures based on the fracture energy release are implemented through a Mathematica © notebook reported in Appendix A, where the calculations are explicated for the case of the ballista that shoots large spherical projectiles and *polybalos* [4] that represent an automatic darts shooter that is able to throw many little arrows having a mass of 150 g.

The results compared with the finite element model and with the literature outfit show that the simplified approach, although very little time-consuming and requiring minimal memory allocation, it is able to offer good results and allows analyzing, with minimal effort, the effect of projectiles on the wall staring on the measure of the target destroyed volume and mechanical properties, namely the elastic parameters, plastic stress limit, and the fracture toughness obtained using the material stress intensity factor.

Author Contributions: Conceptualization, V.M., S.P. and R.Z.; methodology, V.M., S.P. and R.Z.; software, V.M.; validation, V.M., S.P. and R.Z.; formal analysis, V.M., S.P. and R.Z.; investigation, V.M., S.P. and R.Z.; resources, V.M., S.P. and R.Z.; data curation, V.M., S.P. and R.Z.; writing—original draft preparation, V.M., S.P. and R.Z.; writing—review and editing, V.M., S.P. and R.Z.; visualization, V.M., S.P. and R.Z.; supervision, V.M.; project administration, V.M.; funding acquisition, V.M. All authors have read and agreed to the published version of the manuscript.

Funding: This research was supported by the project "SCORPiò-NIDI", CUP B53D2302210 0006, funded by the Italian Ministry of Research under the PRIN (DD n.104/2022) funding initiative.

Institutional Review Board Statement: Not applicable.

Informed Consent Statement: Not applicable.

Data Availability Statement: Research data are available upon request.

Acknowledgments: The author thanks Gabriel Zuchtriegel, Director of the Archaeological Area of Pompeii, Giuseppe Scarpati, Head of the Study and Research Area, and Valeria Amoretti. The research activities are part of the MUR—PRIN 2022 project "SCORPiò-NIDI".

Conflicts of Interest: The authors declare no conflicts of interest. The funders had no role in the design of the study; in the collection, analyses, or interpretation of data; in the writing of the manuscript; or in the decision to publish the results.

Appendix A

In the appendix, a notebook for Mathematica © Wolfram Research Inc. calculates the velocity of the projectile from the hole track dimensions using fracture energy release.

The notebook consists of two modules calculating the velocity of the projectile in the case of the spherical projectile from a ballista and the pyramidal projectile from a *polybolos*.

```
Mathematica (C) notebook for calculating projectile momentum

(* Initialization of units*)

mm = 0.001 m;
J = New m;
New = kg m / s^2;
M = 1000 000;
Tough = 0.00081 M J / m^2;

In[ ]:= (* Case of the sphere projectile *)

In[ ]:= vP[df_] := Module[{DSphere = 140 mm,
      HoleDept = 120 mm, dFrag = df mm},
    BallVol = 4 / 3 Pi (DSphere / 2) ^3;
    HoleVol = Pi HoleDept^2 (DSphere / 2 - HoleDept / 3);
    BallMass = 2800. BallVol kg / m^3;
    aico = Sqrt[0.5 - Sqrt[5.] / 10] dFrag;
  (*FragVol=4/3 Pi (dFrag/2)^3; sphere*)
    FragVol = 5 / 12 (3 + Sqrt[5.]) aico^3; (*icosahedron*)
    nfrag = HoleVol / FragVol;
    (*FragSurf=4 Pi (dFrag/2)^2; sphere*)
  FragSurf = 5 Sqrt[3.] aico^2;
  FractSurf = nfrag * FragSurf Sqrt[5.] / 2;
  FractEnergys = Tough FractSurf;
  ballveloc = v /. NSolve[1 / 2 BallMass v^2 == FractEnergys, v][[2]];
  BallSpeed = PowerExpand[ballveloc /. New → kg m / s^2];
  BallSpeed /. {m → 0.001 km, s → 1 / 3600 hr}]

In[ ]:= (* Case of the darts*)

In[ ]:= vd[df_] := Module[{B = 30. mm,
      H = 40. mm,
      h = 0.0000001 mm,
      dFrag = df mm},
    b = B / (H + h) * h;
    HoleVol = B^2 * H / 3 - b^2 h / 3;
    DartMass = 0.15 kg;
    aico = Sqrt[0.5 - Sqrt[5.] / 10] dFrag;
    (*FragVol=4/3 Pi (dFrag/2)^3; sphere*)
    FragVol = 5 / 12 (3 + Sqrt[5.]) aico^3; (*icosahedron*)
    nfrag = HoleVol / FragVol;
  (*FragSurf=4 Pi (dFrag/2)^2;*)
    FragSurf = 5 Sqrt[3.] aico^2;
  FractSurf = nfrag * FragSurf Sqrt[5.] / 2;
  FractEnergyd = Tough FractSurf // Simplify;
  vdart = v /. NSolve[1 / 2 DartMass v^2 == FractEnergyd, v][[2]];
  vdart /. New → kg m / s^2;
  DartSpeed = PowerExpand[vdart /. New → kg m / s^2];
  DartSpeed /. {m → 0.001 km, s → 1 / 3600 hr}]
```

References

1. Maiuri, A. Studi e richerche sulla fortificazione di Pompeii. *Monum. Antichi Pubblicati Per Cura Della Reale Accademia dei Lincei* **1929**, *33*, 113–290.
2. Russo, F.; Russo, F. *89 a.C.: Assedio a Pompei: La dinamica e le tecnologie belliche della conquista sillana di Pompei*; Flavius Edizioni: Rome, Italy, 2005.
3. Thakkar, M.M.; Ardeshiri, A.; Guagliano, M. Structural Integrity Assessment of Pompeii's City Wall under Roman Artillery Fire: A Finite Element Approach. In *Discovering Pompeii: From Effects to Causes. From Surveying to the Reconstructions of Ballistae and Scorpiones*; (under process of publication); Realcasa dell'Annunziata, Department of Engineering, Vanvitelli University: Naples, Italy, 2025.
4. Rossi, A.; Gonizzi Barsanti, S.; Bertacchi, S. Use of *Polybolos* on the City Walls of Ancient Pompeii: Assessment on the Anthropic Cavities. *Nexus Netw. J.* **2025**, *27*, 243–272. [CrossRef]
5. Rossi, A. The Survey of the Ballistic Imprints for a Renewed Image of Unearthed Pompeii. *Nexus Netw. J.* **2024**, *26*, 307–324. [CrossRef]
6. Bertacchi, S.; Gonizzi Barsanti, S.; Rossi, A. Geometry of Wall Degradation: Measuring and Visualising Impact Craters in the Northern Walls of Pompeii. *SCIRES-IT-Sci. Res. Inf. Technol.* **2024**, *14*, 111–128. [CrossRef]
7. Rossi, A.; Bertacchi, S.; Formicola, C.; Gonizzi Barsanti, S. Piccole indentazioni antropiche rinvenute nella riesumata cinta urbica di Cornelia Veneria Pompeii. *Disegnare IDEE Immagin.* **2025**, *69*, 54–67.

8. Rossi, A.; Gonizzi Barsanti, S.; Bertacchi, S. Natural or anthropic? Measurement and visualisation of wall cavities in city walls. In *Transitions, Proceedings of the 45th International Conference of Representation Disciplines Teachers*; Franco Angeli: Milano, Itlay, 2024; pp. 1957–1978, Measure/Out of Measure. [CrossRef]
9. Balme, M.R.; Rocchi, V.; Jones, C.; Sammonds, P.R.; Meredith, P.G.; Boon, S. Fracture toughness measurements on igneous rocks using a high-pressure, high-temperature rock fracture mechanics cell. *J. Volcanol. Geotherm. Res.* **2004**, *132*, 159–172. [CrossRef]
10. Cattania, C.; Rivalta, E.; Hainzl, S.; Passarelli, L.; Aoki, Y. A nonplanar slow rupture episode during the 2000 Miyakejima dike intrusion. *J. Geophys. Res. Solid Earth* **2017**, *122*, 2054–2068. [CrossRef]
11. Liu, K.; Zhao, J. Progressive damage behaviours of triaxially confined rocks under multiple dynamic loads. *Rock Mech. Rock Eng.* **2021**, *54*, 573–590. [CrossRef]
12. Minutolo, V.; Gesualdo, A.; Nunziante, L. Local Collapse in Soft Rock Bank Cavities. *J. Geotech. Geoenvironmental Eng.* **2001**, *127*, 1037–1042. [CrossRef]
13. Heap, M.J.; Violay, M.E. The mechanical behaviour and failure modes of volcanic rocks: A review. *Bull. Volcanol.* **2021**, *83*, 33. [CrossRef]
14. Zona, R.; Ferla, P.; Minutolo, V. Limit analysis of conical and parabolic domes based on semi-analytical solution. *J. Build. Eng.* **2021**, *44*, 103271. [CrossRef]
15. Zona, R.; Minutolo, V. A dislocation-based finite element method for plastic collapse assessment in solid mechanics. *Arch. Appl. Mech.* **2024**, *94*, 1531–1552. [CrossRef]
16. Pei, G.; Xiao, D.; Zhang, M.; Jiang, J.; Xie, J.; Li, X.; Guo, J. Study on the Dynamic Fracture Properties of Defective Basalt Fiber Concrete Materials Under a Freeze–Thaw Environment. *Materials* **2024**, *17*, 6275. [CrossRef] [PubMed]
17. Palladino, S.; Esposito, L.; Ferla, P.; Zona, R.; Minutolo, V. Functionally Graded Plate Fracture Analysis Using the Field Boundary Element Method. *Appl. Sci.* **2021**, *11*, 88465. [CrossRef]
18. Esposito, L.; Palladino, S.; Minutolo, V. An effective free-meshing and linear step-wise procedure to predict crack initiation and propagation. *Theor. Appl. Fract. Mech.* **2023**, *130*, 104240. [CrossRef]
19. Palladino, S.; Minutolo, V.; Esposito, L. Hybrid semi-analytical calculation of the stress intensity factor for heterogeneous and functionally graded plates. *Eng. Fract. Mech.* **2022**, *274*, 108763. [CrossRef]

engineering proceedings

MDPI

Proceeding Paper

Re-Construction of the Small Xanten-Wardt Dart Launcher †

Michele Fratino [1], Luis Palmero Iglesias [2] and Adriana Rossi [3,*]

1 JustMO' Cultural and Creative Enterprise, 86100 Campobasso, Italy; michele.fratino@justmo.org
2 Departamento de Construcciones Arquitectónicas, Universitat Politècnica de València, 46022 Valencia, Spain; lpalmero@csa.upv.es
3 Dipartimento di Ingegneria, Università degli Studi della Campania Luigi Vanvitelli, 81031 Aversa, Italy
* Correspondence: adriana.rossi@unicampania.it
† Presented at the Conference "Discovering Pompeii: From Effects to Causes—From Surveying to the Reconstructions of Ballistae and Scorpiones", Aversa, Italy, 27 February 2025.

Abstract: Based on the dimensions of the small Xanten catapult, this study reconstructs a full-scale model to validate its manufacturing techniques and evaluate its effectiveness. The process underscores the role of experimental archaeology: the activity facilitates a dynamic sequence of queries, guides the interpretation of signs—not merely physical ones—refines the perception of the cognitive model, and relies on an interdisciplinary approach and strategy. The reconstruction fosters social engagement and scientific dialogue, supporting the adoption of new strategies for knowledge transmission and cultural valorization. The conclusions of this study contribute to the debate on the causes of damage inflicted by the Roman legions on the perimeter walls of Pompeii.

Keywords: experimental archaeology; Pompeii siege; Roman artillery; Sullan siege; terminal ballistics; Xanten-Wardt; Ampurias; handheld scorpion weapon; ancient weapons reconstruction; scorpion dart launcher; full-scale replica

1. Introduction

There are several reconstructions of the Xanten-Wardt, a small dart launcher from the Imperial era (1st century AD) discovered in 1999 at the bottom of a river, now Lake Südsee, in the administrative district of Düsseldorf (Germany). The minor tributary of the Rhine passed through *Castra Vetera* [1], a legionary camp of the Roman province of Germania not far from the town, colonized by the emperor Trajan and consecrated to Saints Vittore and Malloso, from which the toponym *ad san(c)tos*, later became Xanten [2]. The find, imprisoned in the solidified sand on the bottom of the river, has become famous for the use of sophisticated identification techniques (such as computed tomography and X-rays). It is one of the rare examples of a catapult preserved with an almost intact wooden frame. The "*capitulum anatonum,*" in the definition by Vitruvius—who dedicates Book X of *De Architectura* [3,4] to the construction of Roman artillery—in fact is missing only the lateral upright; however, this is an opportunity, rather than a defect, allowing one to view the interior where the pair of elastic torsion bundles was housed. After its restoration [5] the engine has come to be exhibited at the RömerMuseum and Archäologischer Park Xanten (APX) in North Rhine-Westphalia (LVR) (Figure 1).

Tatters of filaments stuck on the bars that hold the bundles to the metal flanges aligned with the upper and lower holes of the frame, observed under an electron microscope, reveal the nature of the connective tissue typical of bovine tendons, which is suitable for transmitting the force accumulated by torsion when the bowstring is released. At first glance, the small size of the wooden frame is striking. Compared with the average sizes of powerful missile launchers that existed in the ancient western world, the diameter of the holes drilled in the frame through which the twisted bundles pass, blocked by a bar to the metal flanges, is almost halved (45 mm). According to Vitruvius, the minimum diameter of the hole drilled in the frame is 79 mm (*De Arch.*, X, 9, 401–405). The flanges found at Emporion (now Ampurias, Spain) have similar dimensions (Figure 2a,b). Based on these measurements, Erwin Adelbert Schramm (1856–1935) reconstructed the first full-scale replica in 1912 (Saalburg Museum), and several functional reconstructions exist today (Figure 3).

Figure 1. The Xanten-Wardt dart launcher frame after conservation in the Römer Museum and Archäologischer Park Xanten. Photograph by dr. Rogier Kalkers, taken on 11 November 2023.

(**a**)

(**b**)

Figure 2. (**a**) Artifacts unearthed at Ampurias (Spain); (**b**) full-scale reconstruction of the Ampurias catapult by Flavio Russo, Archeotecnica.com.

This measurement [6,7] does not differ significantly from the other diameters found during the excavations conducted between 1984 and 1995 in Teruel (near Caminreal), and subsequent excavations in Cremona (Figure 3). It is logical to deduce that the optimal calibers of ancient light artillery were unified, so that worn or broken components could be easily replaced [8]. This necessity was well understood by the most famous scientists and mathematicians of the time. All, directly, reflect the proportional relationship linking the diameter of the hole, taken as a module, and the size of the individual components measured against the whole. The calculation of symmetries described by Vitruvius is exemplary. For the military engineer following Caesar, the typical size of the scorpion, approximately 33 modules in length, 13 modules width, 14 modules at the base (Figure 4), was linked to the weight of the projectiles. Vitruvius reports the correspondences expressed in Roman pounds and fingers (*De Arch.*, X, 10–13, 19 AD).

Figure 3. (**a**) At the top: (**left**) Schramm's reconstruction of the Ampurias frame, Saalburg Museum, and (**right**) the Caminreal frame, reconstructed by Baatz, Aalen Museum, photo by Len Morgan, from [9] (Figure 3). (**b**) In the center: (**left**) Full-scale reconstruction of the Ampurias catapult by Flavio Russo, Archeotecnica.com and (**right**) a sculpture of a catapult on the tombstone of Caius Vedennius Moderatus (81–96 AD), "architect of war machines," found along the Via Nomentana. (**c**) The full-scale reconstruction of the Xanten catapult by Flavio Russo, Archeotecnica.com.

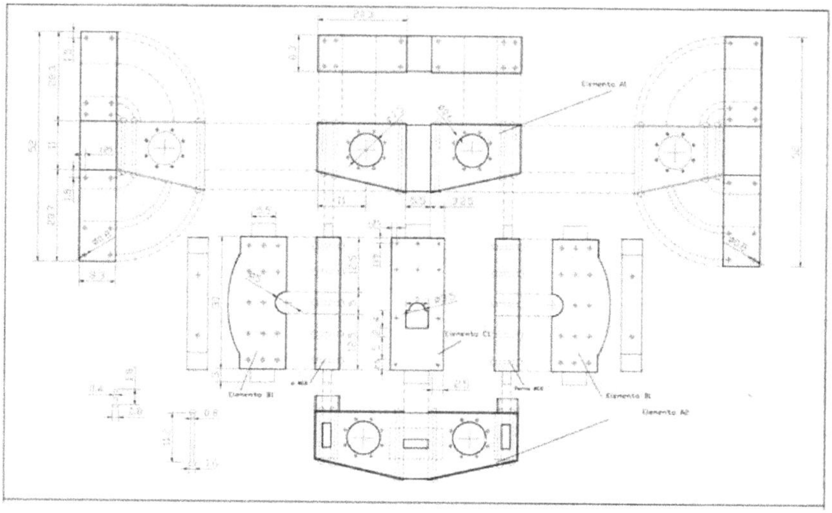

(**a**)

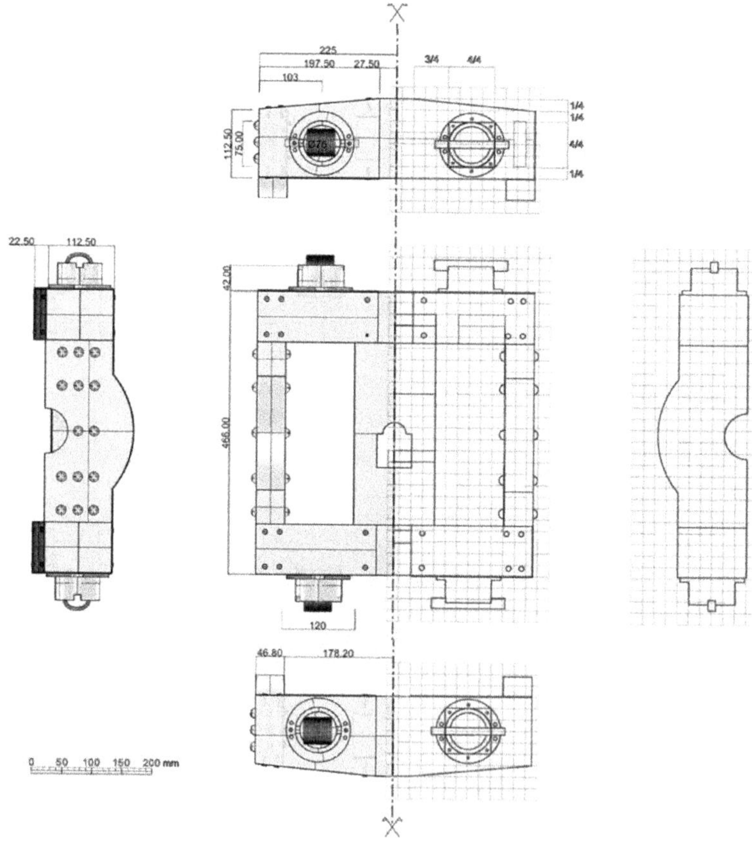

(**b**)

Figure 4. (**a**) At the top: Executive plan of the elastic torsion engine unit—wooden frame; measurements are in centimeters (by A.R.). (**b**) At the bottom: Proportional sizing of the frame according to Vitruvian measurements (by a student, course by Prof. A.R.).

The smallest or largest elastic torsion engines are measured based on these data; the measurements of other components of the system, such as the trigger device, did not vary in proportion to the calibers, being dictated by ergonomic considerations.

It is thus evident that the small Xanten scorpion was not an exception. Flanges of the same size have been found both inside a Tunisian wreck of the Republican era (no. 3, 45 mm, Mahdia, Tunisia) and in the *Volubilis* catapult of the Imperial era (no. 467, 44 mm, Mauritania, Morocco). The *modiolus* of the wooden engine found in 1998–2001 during excavations inside the military laboratory and depot near the hill fortresses of Carlisle Castle is also of similar diameter. Alan Wilkins has visually compared the proportions between the reconstruction of the small Xanten scorpion directed by him and the find now on display at the Tullie House Museum [9]. The time span that covers the findings mentioned above confirms the standardized use of catapults, even if of small caliber, as they are easy to handle, dismantle and transport. The authors were also fascinated by these features. They therefore chose to build a full-scale demonstrator, calibrated via the documented measurements, to encourage an interactive dialogue with scholars and students on the wise and sophisticated engineering of Imperial–Republican era artillery. This is a theme around which comparative analyses and certified reconstructions of ballistae and scorpions revolves. The research was funded by the MUR because it is considered original and innovative. The construction phases, almost as if they were moments of an archaeological investigation, follow the steps necessary to create an experimental product. The question that the article aims to answer, focusing on the time gap, is the technical characteristics of the weapon, those on which one can base the reconfiguration of terminal ballistics parameters or, in other words, those characteristics that produced the damage that can be seen in the present day on the extrados of Pompeii's city walls [10,11].

2. Materials and Methods

In his treatise on siege machines for demolitions at a distance (βελοποιικά) Philo of Byzantium, active between 280 and 220 BC, proportions the ballistae based on the *modiolus*: "The diameter of the opening that passes the turnbuckle [the elastic bundle, ed.] is the starting point" [8] (p. 193).

It is therefore necessary to define the diameter of the hole and the module to derive the components of dart and stone throwers. None of the treatise writers, however, "indicates the convenient way nor the procedures for the construction of the war machines, nor the way to employ them in a satisfactory manner, it being the habit of authors to write for readers whom they supposed to be already informed about all the details" [8] (p. 173). The contribution of Vitruvius's treatise (*De Arch.*, X, 9, 1–4, 401–405) is therefore important and should be complemented with the later treatise on machines attributed to the genius of Heron of Alexandria the elder (1st century AD) and transcribed from ancient Greek into Latin [12,13]. Systematically ignored, the volume, translated from ancient Greek into Latin, was rediscovered by H. Kochly and W. Rustow, who reintroduced the elastically propelled artillery treatises by printing the *Griechische Kriegsschriftsteller,* published in Leipzig in 1853 [7] (p. 30). Real studies began only after the discovery in Lyon, between 1855 and 1987, of a metal box which some assigned as belonging to the driving body of the Heron's *cheiroballistra* [14].

The find stimulated studies by enthusiastic scholars [15], who indirectly made understandable the meaning of a kind of axonometric exploded view contained in some folios of the *Codex Parisinus inter supplementa Greca* 607, foll. 56r-58v, Bibliothèque Nationale, Paris [16] (pp. 201–209). A young army officer, Erwin Adelbert Schramm (1856–1935), settled the controversy surrounding the functioning of the Roman artillery [17]. Based on the interpretation of Philo's Greek text (βελοποιικά) Schramm, then an army major, built a demonstrator that was tested in front of the Kaiser (1912). The tables attached to his German translation describe to scale the mechanisms [17] (cf. Tab. 7) were studied in depth in the following period, translated into English [6] and repeatedly commented upon [18,19]. The original fragments of Philo's text are annotated by Flavio Russo, historical advisor to the army staff. Among the working demonstrators reconstructed by the scholar are the scorpion and ballista exhibited at the Archaeological *Area of Saepinum*, Altilia (CB), (Figure 3b) and the small Xanten scorpion (Figures 3c and 5). Flavio Russo's theoretical and practical work is the basis for the life-size reconstruction carried out for SCORPiò-NIDI by Dr. Michele Fratino, archaeologist and member of the cultural and creative enterprise JustMO'.

(**a**) (**b**)

Figure 5. (**a**) Front and (**b**) back views of the Xanten engine by Flavio Russo. Courtesy of the author.

3. Results: Reconstruction of Xanten

It is difficult for us to ascertain to which inventor to attribute the appearance of the propeller sculpted on the tombstone of Caius Vedennius Moderatus (81–96 AD), "architect of war machines" as explained in the slab found along the Via Nomentana in 1816. On the other hand, the formal analogy between the machine carved on the tombstone and the view of the small engine exhibited at the museum of the Xanten Archaeological Park is quite evident (Figure 3b).

The restoration offers scholars the wooden frame, and offers it in an excellent state of conservation for further study. According to the indications provided by Philo of Byzantium, the machine is of the *euthyton* type: the arms make an "effort in the right direction" (*eu-tònos* in Greek) to twist the pair of filaments of the bundles. The reverse movement of the arms (*palin-tònos* in Greek) allows an excursion greater than 90°, giving, with the same module (*tònos*), greater power to the machine. Even if the dimensions are proportionally decreased, the following objective is achieved: "to launch a missile at a long distance on a given target to deliver a powerful blow" [7]. Although it is a *euthyton*, the dart launcher found in Xanten draws greater power from the use of iron arms that replace wooden ones. The greater effort to which the frame is subjected is supported by the armor plating that reinforces it (Figure 5).

The module is, as usual (Vitruvius, *De Arch.*, X, 10–13, 19 AD), derived from the diameters of the holes drilled in the wooden frame, a detail precisely recorded in the technical survey drawings of the artefact.

An accurate reconstruction of the small scorpion was carried out by engineers Len Morgan and Tom Feeley under the guidance of Alan Wilkins, an expert in ancient history and archaeology. Their reconstruction is a faithful replica: millimetric precision reproduces construction flaws, while highlighting the functional defects (Figure 6a) [20].

(**a**) (**b**)

Figure 6. (**a**) The mistake in drilling the spring-holes off-centre, which leads to spring-rope wear [9] (Figure 4). (**b**) Reconstruction of the small scorpion by Michele Fratino for SCORPiò-NIDI.

Our demonstrator, by contrast, while based, as is appropriate [21], on the survey of the torsion elastic engine, is primarily aimed at verifying the construction procedures, the use of techniques, and the materials compatible with the period. The analysis is therefore focused on the functioning of the components, such as the shooting and loading/unloading devices of the bowstring, nowadays entrusted to the ingenuity of individuals, as the ancient treatise writers took it for granted that their readers already knew all of the necessary details [8] (p. 173). In the case of the small scorpion, a question to be resolved is the absence of winches, essential for the use of stationary catapults.

Our demonstrator, however, although based, as is necessary for experimental archaeology [22], on material, literary, iconographic sources, intends to verify the functioning of the components and especially those of which there is no mention or description in the ancient treatises. It is implicit, for example, that the "handheld" catapults do not provide winches, which is essential instead for the ground ones. However, despite the small caliber, they had to be loaded with the aid of a mechanism, human strength not being sufficient for the purpose, something which in turn caused the abandonment of the previous *gastraphete* [8]. The issue was resolved using a ratchet with the aid of a support. The balance point of the weapon in the hand is immediately behind the frame [23]. The endless harpoon and the release device were integrated through comparative study, ensuring that the choices were consistent with the contextual manufacturing criteria and methods of the reference culture (Figure 6a,b).

Our demonstrator replicates the loading device applied by Russo [21]: the mechanical ratchet consists of a toothed wheel, a beak or coplanar tooth that allows movement in only one direction, except for a certain mechanical game related to the distance between the teeth. The teeth of the wheel are asymmetrical and angled so that there is little friction in the sliding direction, and vice versa in the locking direction, the tooth mates with the beak to create a large surface to oppose the movement of the wheel. The orientation of the parts is fixed a priori, a rigid and hinged beak can exploit the force of gravity to move into the locking position (Figure 7).

(a)

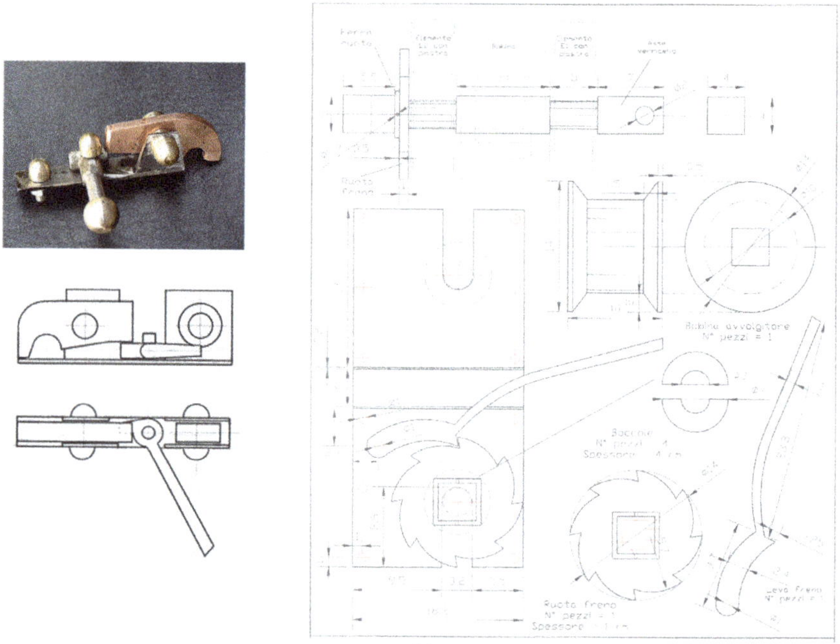

(b)

Figure 7. (**a**) Loading of the small scorpion. (**b**) Left: Detail of the ratchet. Right: Technical drawings—plate, axle, winch, wheel, and lever (drawings by A. R., for *Archeotecnica* by F. Russo).

Particular attention was paid to the construction of the projectile. For catapults, Vitruvius specifies, drawing on Philo, that «All the proportional relationships of the organs of these weapons are calculated on the basis of the length of the dart they have to hurl, the ninth part of which is made to correspond to the hole in the frame, through which passes the bundle of twisted fibers that support the arms» (Vitruvius, *De Arch.*, X, 9, 1–4, 401–405). Knowing the diameter of the *modiolus* (Figure 8a,b), we can hypothesize the length of the dart according to Heron's calculation [8], $d = L/9$. The tip, made of iron and rarely bronze, must be pointed and sufficiently long to penetrate adequately. Forged on the anvil, it was vaguely pyramidal. However, there is no lack of conical tips. Among those preserved in museums, some are inserted into the wooden shaft with cannon-shaped or tang-shaped finishes [24]. Compared with the whole, the metal tip represented a significant fraction of the total weight of the dart. To keep the projectile in the right balance, experience teaches that the center of mass must be positioned at about 1/5 of the total length. To stabilize the trajectory in the stretched shot part, they were equipped with feathers. It is undoubtedly emblematic that Ammianus Marcellinus, a Roman historian and soldier of the late imperial age, mentions in their reports the horizontal position of two feathers, with a third placed vertically upwards to avoid their rubbing along the launch channel [8].

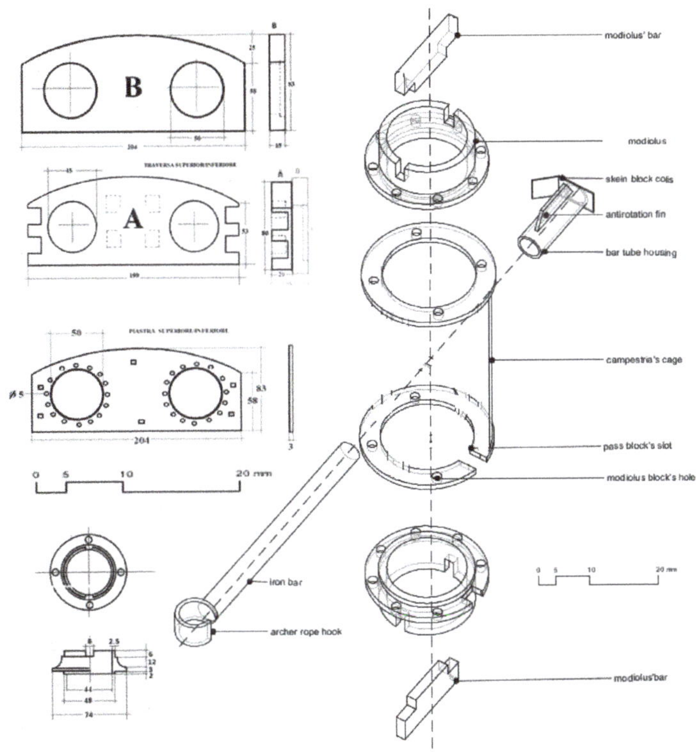

(a)

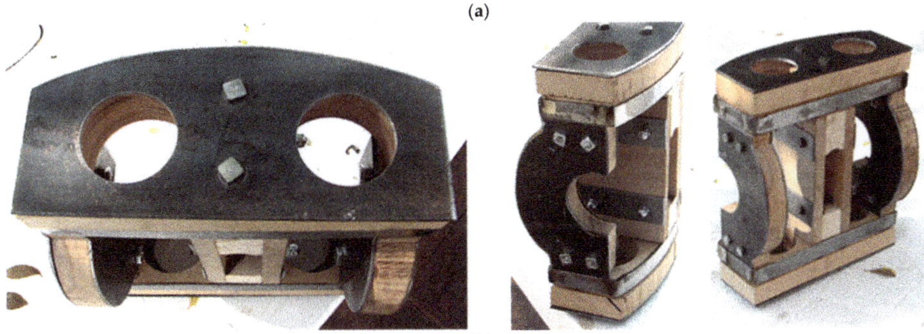

(b)

Figure 8. (**a**) Technical drawings of the upper and lower crossbars of the frame of the Xanten modiolus: on the left, the plan and an elevation view; on the right, an axonometric exploded view. (**b**) The Xanten engine under construction by Flavio Russo.

Test

The idea of recording the trajectory ranges to inductively calculate the speed of the dart in relation to the distance covered is supported by elementary knowledge that current digital techniques (e.g., Sabre Sky Screen) make reliable. Accurate measurements are obtained by connecting a photoelectric device to the PC; through the light decrease of the 3D digital timer sensor one can record the passage of the projectile on the screen. Modern equipment for ballistic testing also allows for testing parameters that are useful for calculating terminal ballistics [25,26]. However, the use of a refined technology is not decisive for our interphase objective. The shooting carried out with a mobile device allowed for the capture of 10 frames in a total of one minute. Taking the fixed ground supports as reference, and using the known length of the dart, the displacement in each frame is calculated (Figure 9).

(**a**)

(**b**)

Figure 9. Dart throwing test. (**a**) First and last frames and (**b**) distance covered, showing the dart's position in each frame.

The approximations generated by the perspective distortion, the blurring generated by the scale enlargement, the bending of the arrow and the considerably unstable shot—a result of its placement without support—are evident. Nonetheless, the experiment demonstrates the validity of the approach that allows us to trace the causes from the effects. From the first position of the arrow visible in its length (as recorded in the third frame) the dart would move about 67 cm (Table 1). Converting the distance into meters and the time into seconds, a speed of about 20.65 m/s (i.e., 3.442 m/0.1667 s) is obtained. A negligible quantity compared with that derived and the laws of terminal ballistics formalized in the 1st century AD through the "patient repetition of the result" [8] and theoretically applied to the ballistic effects detected along the urban circle of Pompeii (cf. Russo and Rossi article in these proceedings).

Table 1. Correspondence between the 10 images (60 fps each) and the distance covered by the dart in centimeters.

FRAME		Distance Covered in cm
1	2	
2	3	
3	4	33.6
4	5	35.8
5	6	35.3
6	7	34.3
7	8	36.1
8	9	33.8
9	10	35.4
10	11	35.1
1	11	344.2

To perform "safe" shots, considering the group of students present, the tension of the bundles was reduced. But the real reason for the low speed compared with the real power of the catapults lies in the elasticity coefficient of the hemp rope, suitable for the construction of a demonstrator for exhibition but certainly not suitable for shooting tests. The shreds of connective tissue that remained stuck and examined under the microscope for Xanten, suggest the use of animal nerve fibers. Their elasticity is today approximated with the production of synthetic fibers intended for the reconstruction of intra or extra-articular ligaments. It would therefore be necessary to repeat the experiment with suitable ropes, perhaps those used and mentioned above. Our construction is therefore neither the first nor the best, but was useful to retrace thc historical and archaeological data. It was important to approach construction phases, verify the functioning, and demonstrate the usefulness of a "cognitive gaze" that is common to the tactile and optical experience in order to read the concrete and invisible signs of what was collected and integrated.

4. Discussion

Experimental archaeology is a discipline that has been successfully established in the European panorama in order to study, disseminate and enhance the inherited heritage. Re-enactment and living history follow experiences oriented towards the verification, reproduction, communication and simulation that benefit archaeological sites, especially open-air museums [27]. The idea of relaunching the image of the city of Pompeii, disseminating the experimental activity and the historical–military study carried out within the MUR-funded research, applies the experimental method and provides an interpretation of the archaeological data based on hypotheses considered valid because they are refutable [28].

The reconstruction of the small handheld scorpion contributes to knowledge through a dynamic process of questioning that began with the speculations of Schramm (1856–1935). The demonstrator tested in front of the Kaiser proved to put an end to the disputes raised through the interpretation of the cryptic Greek text (*βελοποιικά*). About thirty *modioli*, half a dozen ratchet stops, six supports for coils of various size; a *kamarion* and several iron and bronze armor plates for catapults and ballistae, a frontal shield and some winch fragments are the finds on which Flavio Russo's study draws [8,21]. The procedural phases of the 1:1

scale reconstruction, carried out by Michele Fratino for SCORPiò-NIDI, refer to his theoretical and practical work. The undocumented parts were integrated through comparative study. Constant references are the texts by Alan Wilkins and the reconstructions by Len Morgan and Tom Feeley. The practice of building physical models [29] makes the heuristic aspect tangible by proposing multidirectional uses in which the "phenomenized" drawing dispels doubts within a unitary dialectical, cyclical and interactive path. Brunelleschi used these to convince patrons fascinated by the realistic views of the painters [30] (p. 117); Filarete considered them an erudite homage to their clients [31] (40; 207); Alberti used them to verify the calculation of the "symmetries" [32] (I: 860–862); Michelangelo built them to provide construction sites with a safe guide [33]; and Vasari used them to study precision [34] (V: 18; 467) and the very small and the very large.

All objectives are not lost despite the digital constructs, the multimedia and multimodal fruition that accompanies the advanced visualization, the printing of digital copies [35] and the sophisticated calculations on the inverse models.

5. Conclusions

The experience gained through Flavio Russo's Archeotecnica.com has allowed us to document, almost as if they were moments of an archaeological investigation, the steps necessary to create a material product calibrated to the measurements of the find discovered in the vicinity of Xanten. A dynamic process of investigation supports the experiment. Drawing on documentary and iconographic sources and ancient construction techniques, particular care was taken in choosing the appropriate tools to cut the solid beech wood, beat the iron and forge the bronze parts. The weak point of the full-size demonstrator remains the nature of the filaments twisted by the arms to accumulate elastic energy to be released on command. Despite the evident approximation, the test carried out in the presence of scholars and students demonstrates the usefulness of adopting methods of experimental archaeology. Without forgetting the involvement of the senses and emotions, visitors were transformed into actors, confirming that the best strategy to extend the protection and enhancement of the inherited heritage is their involvement. In situ and online platforms help to amplify people's ability to socialize around objects and physical environments to make "Science" dialogue with the perception of the cognitive model.

These conclusions fuel the debate on the determination of the causes that produced the damage (1st century BC) detected on the extrados of the city walls of Pompeii and attributed to the siege of Sulla, the topic of the current comparison. Merging new forms of knowledge and promoting a concert of science, humanity and humanism is a duty of our era. The connections fall into the so-called Third Mission. Research and teaching are primary institutional activities of the University, which promotes basic research, applied research and the transfer of technological innovation to the economic–social system, contributing to meeting the development needs of society through its scientific and professional skills. Enhancing research results through the creation of brands, patents and industrial spin-offs leads to the development of innovations suitable to satisfy market demand and needs and/or to generate an improvement in environmental and social impact.

Author Contributions: Conceptualization, A.R.; methodology, A.R. and M.F.; reconstruction, M.F.; validation, A.R., M.F. and L.P.I.; formal analysis, A.R.; investigation, A.R.; resources, A.R.; data curation, A.R., M.F. and L.P.I.; writing—original draft preparation, A.R.; writing—review and editing, A.R.; visualization, A.R.; supervision, M.F., L.P.I. and A.R.; project administration, A.R. All authors have read and agreed to the published version of the manuscript.

Funding: This research was partially supported by the project of significant national interest (call DD n. 104/2022), PRIN 22, prot.20222RJE32 18/9/23 SCORPiò-NIDI, CUP B53D23022100006 (DD n. 1012/2023).

Institutional Review Board Statement: Not applicable.

Informed Consent Statement: Not applicable.

Data Availability Statement: The following supporting information can be downloaded at https://www.facebook.com/reel/1114805353731460 (accessed on 10 March 2025); https://www.instagram.com/p/DHsge4Vt-qm/ (accessed on 10 March 2025).

Acknowledgments: The authors would like to thank Flavio Russo for the teachings and materials made available and the patient guidance in studying the topic in depth.

Conflicts of Interest: The authors declare no conflicts of interest.

References

1. Borger, H.; Oedinger, W. *Beiträge Zur Frühgeschichte Des Xantener Viktorstiftes*; Rheinische Ausgrabungen; Rheinland-Verlag: Düsseldorf, Germany, 1969; Volume 6.
2. Hinz, H. *Xanten zur Römerzeit*; Dom-Buchhandlung Verlag: Deutschland, Germany, 1971.
3. Vitruvius, P. *De Architectura*; Gros, P., Ed.; Einaudi: Torino, Italy, 1997.
4. *L'architettura di Marco Vitruvio Pollione Tradotta e Comentata dal Marchese Berardo Galiani*; Galiani, B., Ed.; nella Stamperia Simoniana: Napoli, Italy, 1758.
5. Schalles, H.-J. *Die Frühkaiserzeitliche Manuballista aus Xanten-Wardt*; Xantener Berichte Band; Philipp von Zabern: Mainz, Germany, 2010; ISBN 978-3-8053-4274-2.
6. Marsden, E.W. *Greek and Roman Artillery. Technical Treatises*; Clarendon Press: Oxford, UK, 1971; ISBN 978-0-19-814269-0.
7. Russo, F. *L'artiglieria Delle Legioni Romane. Le Macchine da Guerra che Resero Invincibile L'esercito Romano*; Istituto Poligrafico e Zecca dello Stato: Roma, Italy, 2004; ISBN 978-88-240-3444-9.
8. Gille, B. *Storia Delle Tecniche*; Editori Riuniti: Roma, Italy, 1985; ISBN 978-88-359-2903-1.
9. Wilkins, A. The Xanten-Wardt Roman Torsion Catapult and Catapult Parts from Carlisle 2013. Available online: http://www.romanarmy.net/pdf/The%20Xanten-Wardt%20and%20Carlisle%20catapult%20finds.pdf (accessed on 10 March 2025).
10. Rossi, A.; Gonizzi Barsanti, S.; Bertacchi, S. Use of Polybolos on the City Walls of Ancient Pompeii: Assessment on the Anthropic Cavities. *Nexus Netw. J.* **2025**, *27*, 243–272. [CrossRef]
11. Bertacchi, S.; Gonizzi Barsanti, S.; Rossi, A. Geometry of Wall Degradation: Measuring and Visualising Impact Craters in the Northern Walls of Pompeii. *SCIRES-IT—Sci. Res. Inf. Technol.* **2024**, *14*, 111–128. [CrossRef]
12. Heron, A. *Heronis Mechanici. Liber de Machinis Bellicis, Necnon Liber de Geodaesia*; Apud Franciscum Franciscium Senensem: Venezia, Italy, 1572.
13. Baldi, B. *Heronis Ctesibii Belopoeeca, Hoc Est Telifactiva. Bernardino Baldo ... Illustratore et Interprete; Item Heronis vita Eodem Auctore*; Typis Davidis Franci: Augusta Vindelicorum, Germany, 1616.
14. Prou, V. *La Chirobaliste d'Héron d'Alexandrie. Notices et Extraits des Manuscrits de la Bibliothèque Nationale et Autres Bibliothèques*; Imprimerie Nationale, Belles Lettres: Paris, France, 1877; Volume 26.
15. Schneider, R. Herons Cheiroballistra. In *Mitteilungen des Kaiserlich Deutschen Archaeologischen Instituts*; Roemische Abtheilung; Imperiale Istituto Archeologico Germanico: Roma, Italy, 1906.
16. Baatz, D.; Feugère, M. Éléments d'une Catapulte Romaine Trouvée à Lyon. *Gall. —Fouill. Et Monum. Archéologiques En France Métropolitaine* **1981**, *39*, 201–210. [CrossRef]
17. Diels, H.; Schramm, E.A. *Philons Belopoiika, Viertes Buch der Mechanik*; Akademie der Wissenschaften; in Kommission bei G. Reimer: Berlin, Germany, 1919.
18. Wilkins, A. Scorpio and Cheiroballistra. *J. Rom. Mil. Equip. Stud.* **2000**, *11*, 77–101.
19. Campbell, D.B. *Greek and Roman Artillery 399 BC-AD 363*; New Vanguard series 89; Osprey Publishing Ltd.: Oxford, UK, 2003; ISBN 1-84176-634-8.
20. Wilkins, A. Roman Imperial Artillery: Outranging the Enemies of the Empire. In *Archaeopress Roman Archaeology*, 3rd ed.; Archaeopress Publishing: Oxford, UK, 2024; ISBN 978-1-80327-784-4.
21. Russo, F.; Russo, F. *Gli Scorpioni Della Repubblica. Cenni Storici, Reperti, Tavole Ricostruttive. La Catapulta di Ampurias*; ESA: Torre del Greco, Italy, 2008; ISBN 978-88-95430-09-6.
22. Russo, F. *Tormenta: Venti Secoli di Artiglieria Meccanica*; Stato maggiore dell'esercito-Ufficio storico: Roma, Italy, 2002; ISBN 88-87940-48-7.
23. Gudea, N.; Baatz, D. Teile Spätrömischen Ballistenaus Gornea Und Orşova (Rumänien). *Saalburg-Jahrbuch* **1974**, *31*, 50–72.
24. Manning, W.H. *Catalogue of the Romano-British Iron Tools, Fittings and Weapons in the British Museum*; Published for the Trustees of the British Museum by British Museum Publications: London, UK, 1985; ISBN 0-7141-1370-0.
25. Richardson, T. The Ballistics of the Sling. *R. Armouries Yearb.* **1998**, *3*, 44–49. [CrossRef]

26. Richardson, T. Ballistic Testing of Historical Weapons. *R. Armouries Yearb.* **1998**, *3*, 50–52. [CrossRef]
27. Comis, L. Re-Enactment, Living History: Rapporti Con l'archeologia Sperimentale e i Musei Archeologici All'Aperto. In *Rivivere e Comunicare il Passato. Il Contributo Della Rievocazione Dell'evo Antico al Marketing Museale e Territoriale*; IBC: Bologna, Italy, 2014; pp. 7–16. ISBN 978-88-97281-20-7.
28. Popper, K.R. *The Logic of Scientific Discovery*; Hutchinson: London, UK, 2014; ISBN 978-1-61427-743-9.
29. Scolari, M. L'idea Di Modello. *Eidos Can. Grad. J. Philos.* **1988**, *2*, 16–39.
30. Manetti, A. *Vita di Filippo Brunelleschi. Preceduta da la Novella del Grasso*; De Robertis, D., Tanturli, G., Eds.; Edizioni Il Polifilo: Milano, Italy, 1976.
31. Filarete. *Trattato Di Architettura*; Grassi, L., Ed.; Edizioni Il Polifilo: Milano, Italy, 1972.
32. Alberti, L.B. *L'architettura (De re Aedificatoria)*; Portoghesi, P., Ed.; Edizioni Il Polifilo: Milano, Italy, 1966; ISBN 978-88-7050-101-8.
33. Millon, H.A.; Smyth, C.H. *Michelangelo Architetto. La Facciata di San Lorenzo e la Cupola di San Pietro*; Olivetti: Milano, Italy, 1988.
34. Vasari, G. *Le Vite de' Più Eccellenti Architetti, Pittori et Scultori Italiani*; Colla: Firenze, Italy, 2012.
35. Dawkins, O.; Dennett, A.; Hudson-Smith, A.P. Living with a Digital Twin: Operational Management and Engagement Using IoT and Mixed Realities at UCL's Here East Campus on the Queen Elizabeth Olympic Park. In Proceedings of the 26th Annual GIScience Research UK Conference: GISRUK 2018, Leicester, UK, 17–20 April 2018; GIS Research UK (GISRUK). University of Leicester: Leicester, UK, 2018.

engineering proceedings

MDPI

Proceeding Paper

Beyond the Museum: Virtual and Physical Replicas of Pompeii's Siege Marks †

Filippo Fantini [1] and Silvia Bertacchi [2,*]

[1] Department of Architecture, Alma Mater Studiorum–Università di Bologna, 40136 Bologna, Italy; filippo.fantini2@unibo.it
[2] Department of Engineering, Università degli Studi della Campania Luigi Vanvitelli, 81031 Aversa, Italy
* Correspondence: silvia.bertacchi@unicampania.it
† Presented at the Conference "Discovering Pompeii: From Effects to Causes—From Surveying to the Reconstructions of Ballistae and Scorpiones", Aversa, Italy, 27 February 2025.

Abstract: This study investigates the potential of reality-based 3D digital modeling, acquired for scientific purposes, to enhance the understanding and accessibility of ballistic imprints on Pompeii's city walls. These impact marks, attributed to the Sullan siege of 89 BC, were caused by projectiles launched by Roman elastic torsion weapons. High-resolution models were acquired through integrated 3D survey techniques to create both virtual and physical replicas. These assets enhance museum accessibility, offering interactive digital content and tactile 3D-printed replicas for visually impaired and mobility-restricted visitors. The findings highlight the role of digital heritage in archaeological research, conservation, and public engagement, bridging the gap between academic study and inclusive cultural dissemination.

Keywords: 3D documentation; 3D printing; accessibility; archaeological visualization; ballistic imprints; cultural accessibility; digital heritage; haptic perception; interactive visualization; virtual museum

1. Introduction

The model, referred to as a maquette, is one of the most widespread and ancient forms of architectural representation [1]. Constructed with diverse materials and serving multiple purposes, the model functions as both a representation and a physical replica of the actual structure [2,3]. Historically, this representation fulfilled specific roles, including design prefiguration for clients and facilitating communication among workers during construction [4]. As such, the model encapsulated the fundamental design and architectural concepts in their most refined form [5] (pp. 131–163).

Today, the digital version of the model, like its physical counterpart, offers a wider range of applications, including but not limited to recording reality, whence the term "reality-based", which signifies the creation of a replica that reliably adheres to real-world morphometric data. Especially in the Cultural Heritage domain, three-dimensional (3D) digitization has become an essential tool for recording and analyzing architectural and archaeological artifacts in a rapid, accurate, and efficient way. The obtained high-resolution reality-based models, primarily captured for the record of the geometric features and conservation state of physical artifacts, have also become increasingly pivotal to a vast range of analysis and are recommended as dynamic resources for restoration planning and management. Moreover, they have improved the dissemination strategies for public engagement, even for administration entities, making heritage more accessible to a wider audience and fostering the breaking down of physical and cognitive barriers to cultural assets. The latter aspect is becoming crucial as the potentialities of digital assets can effectively convey even complex concepts and enable interactive and tactile experiences [6], capable of developing an appropriate cognitive process both for conventional visitors as well as for those with reduced or even absent visual and/or mobility abilities. Reflecting on the use of digital models as tools to explore and support dissemination activities is part of a climate of increased sensitivity that institutions in charge of the preservation and dissemination of sites of cultural interest have embraced toward weaker categories, as evidenced by the recent revision of the visitor information plaques, made more inclusive, present in many museums or archaeological sites [7,8].

In this framework, this research aims to explore communication strategies that, starting from highly reliable 3D documentation acquired through active and passive sensing technologies for scientific and research purposes, can effectively convey to the widest and most diverse audience possible the initial results of the documentation of the ballistic imprints on the northern city walls of Pompeii (Figure 1), which archaeologists attribute to the Sullan siege in 89 BC [9–11]. This subject is little known [12,13] but of fundamental importance since it is one of the few existing cases, if not the only one, where it is possible to measure the firepower of elastic torsion weapons of Roman artillery of the 1st century BC from the still visible effect of the impact on stone ashlars, the final goal of the SCORPiò-NIDI project.

Figure 1. Ballistic marks on stone ashlars in the northern city walls near Vesuvio Gate.

2. Materials and Methods

The research on ballistic imprints was carried out within the framework of the SCORPiò-NIDI project, of which the initial results on impact marks proved to be particularly significant [14,15]. These traces, caused by projectiles launched by the besieging forces, are distributed along the fortified section between Vesuvio Gate and Ercolano Gate, with higher concentrations near towers and access points (Figure 2). The difficulty in recognizing them stems from several aspects: (i) they are distributed seemingly at random and appear at different heights on the wall, including areas that are barely visible or inaccessible; (ii) the heterogeneous composition of the wall's materials (varying by area and historical period, including *pappamonte* stone, tuff, Sarno limestone, or lava in *opus incertum*) complicates the identification of ballistic imprints without proper guidance in reading the archaeological context, as they can easily be mistaken for natural cavities in the stone ashlars; and (iii) seasonal weed vegetation growth, which can be more or less abundant depending on the time of year, further obscures their visibility.

Currently, the northern section of the walls is inaccessible to the public, pending the completion of the project to restore the pedestrian and bicycle paths and the logistics of visitor flows. As a result, the extensive outer section remains temporarily closed to visitors. In addition, the site's topography presents further challenges: the terrain is partly steep and lacks accessible pathways in this area for individuals with reduced mobility. Thus, a twofold challenge emerges: on the one hand, an objective issue related to physical accessibility, and on the other, the difficulty of making the subject matter comprehensible and intelligible, even to experts. Addressing both issues requires the adoption of innovative strategies for the communication and enhancement of these historical traces. This involves not only an appropriate museological narrative [16] (pp. 28–37) but also tangible tactile supports capable of making the subject matter more accessible, particularly for users with limited mobility or visual impairments [17]. To achieve this, the research leverages high-resolution digital three-dimensional models captured for scientific documentation.

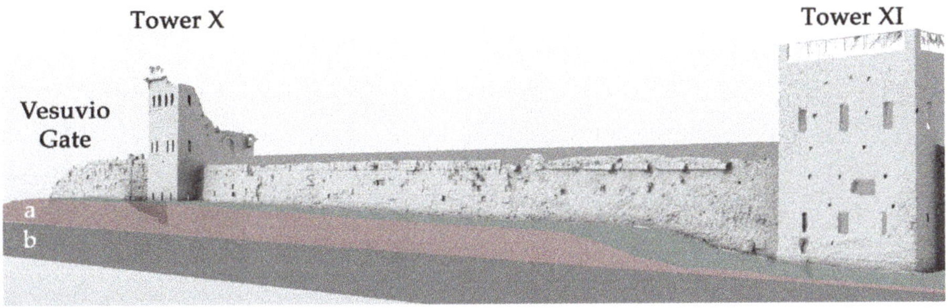

Figure 2. Mesh model of part of the walls near Vesuvio Gate: (**a**) current terrain level; (**b**) probable ancient terrain level at the base of the tower's sally-port door for sorties.

2.1. The 3D Assets

The available 3D models were acquired through an integrated approach, combining multiple techniques with customized pipelines, to achieve the following assets:

1. General documentation of the fortified section, extending approximately 300 m, was conducted using a time-of-flight (ToF) terrestrial laser scanner (TLS) device (Figure 3). This provided a spatial reference framework and facilitated the localization of impact marks, which were more recognizable in sections with regular ashlars;
2. Detailed photogrammetric documentation was carried out using structure-from-motion (SfM) techniques (Figure 4). This focused on specific impact marks, including the deepest and widest ones, caused by stone balls of various diameters (up to 18 cm), as well as smaller traces possibly caused by metal-tipped arrows or lead sling bullets. The research also documented spheroidal stone projectiles still preserved in the museum.

Figure 3. Global point cloud of the surveyed area. Perspective view from Vesuvio Gate toward Tower X.

The integration of multiple survey campaigns and the data processing resulted in high-resolution polygonal models complete with descriptive textures (Figure 5).

Figure 4. An example of detailed photogrammetric documentation of an impact crater case study.

Figure 5. 3D digital model of the remains of Vesuvio Gate and the *Castellum Aquae*: (**a**) Perspective view; (**b**) Orthographic view. Here, the pronounced slope of the cobblestone road, which once led out of Pompeii through the northern gate, is clearly visible.

2.2. The Collection

Based on these 3D high-resolution models, the main purpose was to develop a series of exhibitable assets, enabling site conservators to provide non-expert audiences with effective storytelling tools about ancient siege and defense techniques. These assets serve both museological (scientific-narrative) and exhibitory (perceptual-display) purposes [18].

To this end, the collection includes four main categories of assets, three developed by the authors based on the surveyed data and one created by external experts:

1. Architectural Complex (AC)—A large-scale 3D model of the northern city walls, illustrating building materials, historical construction phases, and the location of case studies [19] (typological and structural aspects) (Figure 6);
2. Architectural Section (AS)—A cutaway model of the city walls including a restored tower and the embankment behind the walls, explaining the spatial organization and proportions between all the elements (proportional aspect) (Figure 7);
3. Architectural Object (AO)—A detailed 3D model of individual case studies, including impact marks and projectiles, to compare the effects of impact on the walls with the blunt objects that caused them [20] (dimensional aspect) (Figure 8);
4. Siege Weapons (SW)—Replicas of ballistae and scorpions, commonly used to launch stone balls and metal-tipped darts, reconstructed at various scales using historical sources [21] (mechanical aspects) (Figure 9).

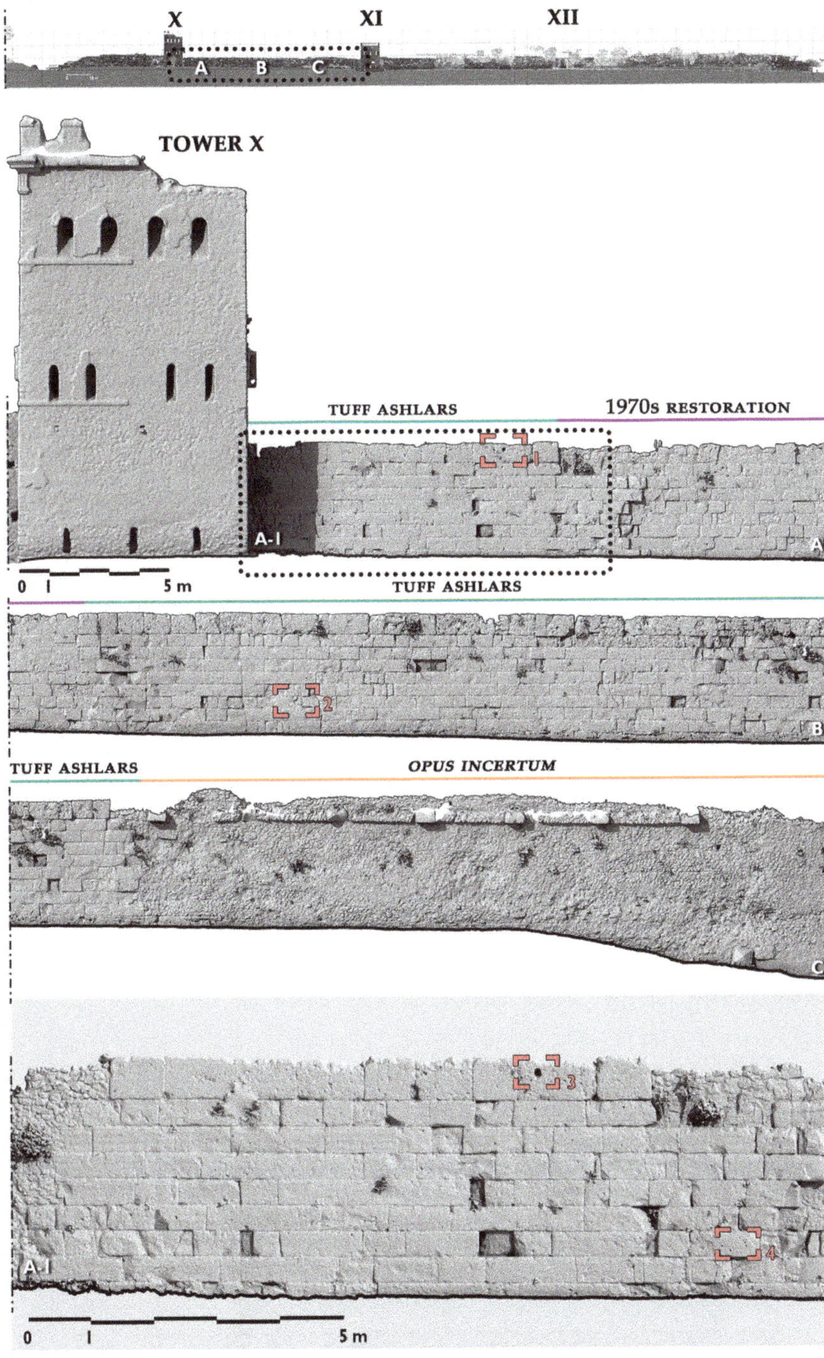

Figure 6. Digital model of the surveyed wall section between the Vesuvio and Ercolano Gates, processed to locate the different types of ballistic traces. Top: General location between the two gates. Middle: Wall sections A, B, and C between Towers X and XI, with masonry-pattern specifications. Bottom: Detailed view of A-1 section.

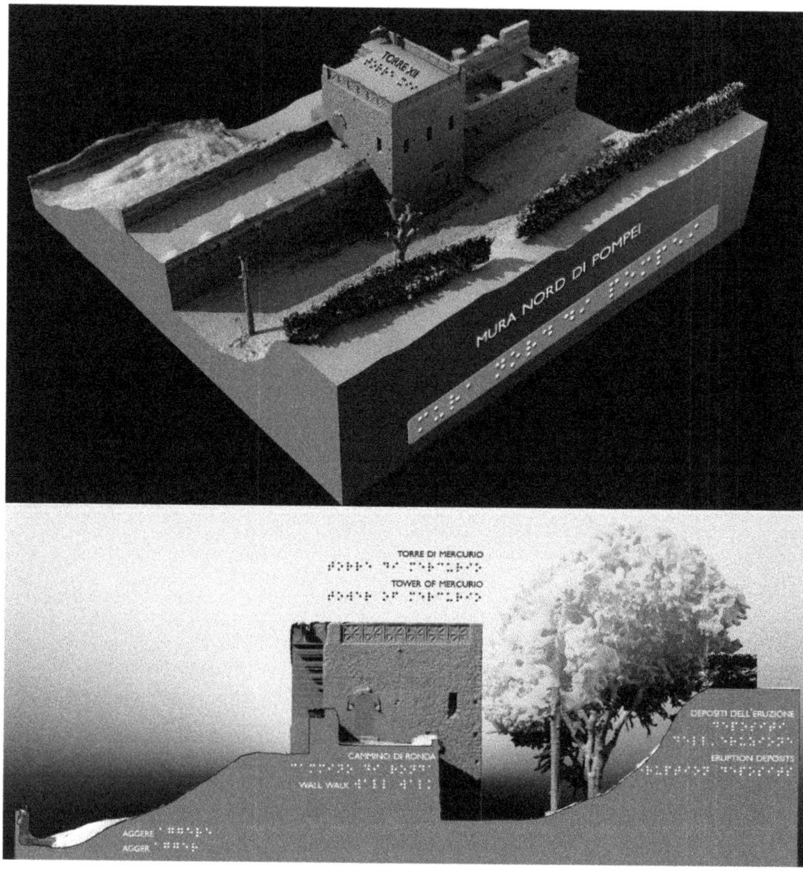

Figure 7. Feasibility study for a 1:50 tactile model of a section of the walls near the Tower of Mercurio, developed from the reality-based digital model to illustrate the spatial articulation of the sloping terrain descending from the northern wall toward Pompeii. Top: Overall layout of the proposed tactile model. Bottom: Detailed section highlighting the unearthed levels with simplified explanatory elements and Braille labels for visually impaired visitors.

Figure 8. Design of the museum installation displaying the 1:1-scale 3D-printed prototypes of wall cavities and stone balls, with reference to the significant wall section, allowing tactile interaction.

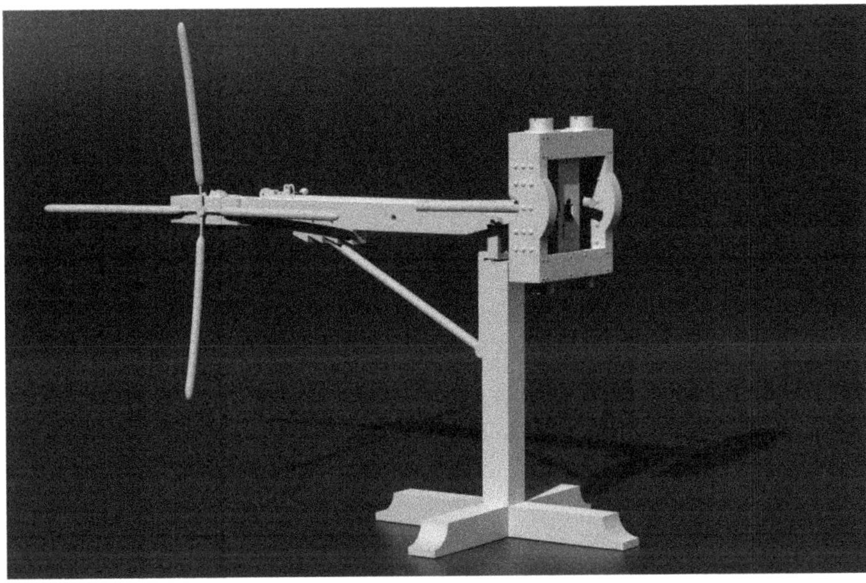

Figure 9. Digital reconstruction of the Roman *scorpio* elastic torsion engine, by C. Formicola.

The developed assets include both virtual content and physical prototypes:

- Virtual content: These digital assets offer high flexibility and can be used for multiple purposes, including professional applications, education, and research. Formats include renderings, orthoimages, simulations, animations, and more. To ensure the most accurate representation, textures have been calibrated to match the apparent color of the original surfaces.
- 3D printed prototypes: Unlike virtual assets, physical replicas require a dedicated design phase to enhance readability and usability following appropriate requirements [22]. Various printing scales were chosen based on artifact type, including 1:100 for AC, 1:50 for AS, and 1:1 for AO.

Additionally, the functioning reconstruction of ancient Roman artillery (SW) was carried out by JustMO', a cultural and creative enterprise. These reconstructions, produced using historically accurate materials, exist at both full scale and reduced scale for educational and experimental purposes.

3. Results

The integration of advanced survey techniques produced a comprehensive and high-resolution 3D dataset that accurately represents the ballistic traces on the northern walls.

The following list details the main steps taken to obtain the different outputs:

- General and detailed digital acquisition phase—The survey approach included both a general measurement based on an active-sensor device due to the large extent of the surveyed area and a series of detailed photogrammetric acquisitions of specific case studies. The 45 medium-resolution TLS scans (10 mm at 10 m) were registered using the Leica Cyclone software (version 9) [23], generating a complete 374-million-point cloud. The photogrammetric survey was conducted with a Nikon D5300 Single-Lens Reflex Camera equipped with a Nikon AF-P 18–55mm f/3.5–5.6 DX VR lens, along with a white balance target to ensure color accuracy.

- Mesh creation—The polygonal mesh processing was carried out using Geomagic Design X software (version 2016.1.1) [24] by manually segmenting the point cloud. This step was necessary due to the large size of the dataset, which was subdivided into different parts based on architectural elements (higher resolution) and other elements such as the environment, vegetation, and connections (lower resolution). The master model, representing the highest resolution version, underwent only minor modifications – such as small hole filling and minimal surface processing—to preserve the original geometry. This version was retained for scientific purposes, such as supporting archaeological hypotheses on the trajectory and impact force of projectiles used during the siege. The photogrammetric models were accurately scaled by aligning control points with the TLS point cloud.
- Optimization phase—From the master model, a lighter version was created for dissemination purposes. This version included geometric reconstructions of missing parts, optimized geometry to reduce file size, and extrapolated missing elements where data were incomplete, using comparative reconstruction techniques (e.g., for the embankment). The goal of this optimized model was to prioritize visual clarity over scientific accuracy, making it more suitable for educational and exhibition purposes [25].
- Texturing—This process was performed using Agisoft Metashape Professional software (version 2.2.1) to reapply the color appearance texture to the final model [26].
- Virtual viewing—3D digital models are typically analyzed by scholars using specialized software with tools that enable detailed analyses, such as geometric pattern recognition, deviation measurement from reference entities, and other advanced functions. However, user-friendly alternatives are now widely available. Increasingly, 3D assets licensed under Creative Commons are accessible through popular online virtual platforms [27], allowing intuitive virtual navigation and interaction via multiple devices. These models can support virtual tours, interactive storytelling, and dynamic exploration. Advanced simulations to better understand the mechanics of ancient artillery through ballistic simulations are already in the planning stages. Additionally, annotations, cultural content, and supplementary materials can be linked to specific parts of the model, enhancing public engagement both on-site and online (Figure 10).
- Design phase—A key step in the development of physical replicas was the selection of a representative section of the city walls. The chosen segment had to both illustrate the main architectural features and offer a clear cross-sectional view of the fortified structure. Additionally, the balance between model detail and printing scale was carefully considered to ensure the best resolution-to-size ratio [28].
- 3D printing phase—Prototype models required watertight geometry and sufficient structural thickness for successful fabrication. Tactile models, designed for haptic exploration, had to meet specific accessibility requirements, including the addition of Braille labels for visually impaired visitors. Particularly, 1:1 scale 3D-printed replicas of impact cavities and projectiles were created to enable a direct comparative analysis between the missile size and the damage caused upon impact (Figure 11). This set of physical replicas, designed for an inclusive museum exhibit, allows visitors to better grasp the scale of the stone projectiles by visually and physically comparing them with the actual damage they inflicted on the walls.

All the proposed applications of digital models have proven to be an effective means of disseminating knowledge on Roman siege weaponry and the ballistic impact marks at Pompeii through emblematic case studies. In particular, tactile replicas of ballistic imprints demonstrated strong potential for accessibility, enabling immersive and interactive experiences that are also accessible to visually impaired visitors.

Figure 10. Virtual view on the Sketchfab platform, model with informative annotations.

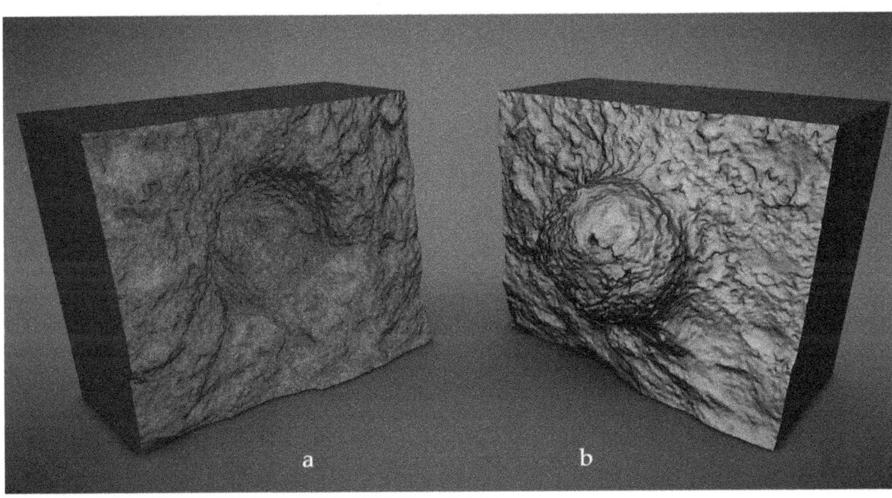

Figure 11. 3D digital model for 1:1-scale replicas: (**a**) positive replica of the ballistic crater in the wall (wireframe enabled); (**b**) its virtual negative mold (Matcap-render visualization).

4. Discussion

The results highlight the significant transformative potential of digital heritage documentation, originally acquired for scientific research, in supporting inclusive dissemination. The integration of digital documentation methods not only enhances the accuracy of archaeological analysis but also improves the accessibility and interpretability of historical findings.

One of the key challenges lies in balancing accuracy and usability. While high-resolution models offer invaluable insights for specialists, their practical implementation in museum settings requires optimization to ensure interactivity and engagement. The integration of digital visualization, 3D prototyping, and tactile exploration represents a promising approach, ensuring that research findings are effectively conveyed to diverse audiences.

Additionally, this study demonstrates how digital reconstructions contribute to heritage conservation. The ability to digitally archive and assess the preservation state of archaeological remains provides a valuable tool for long-term monitoring and preventive conservation strategies.

However, some challenges persist, particularly in effectively communicating the ballistic imprints to general audiences and ensuring that their historical significance is properly understood. For this reason, future initiatives are planned, such as the installation of explanatory panels both on-site and within the museum, enabling visitors to directly compare contextual information with the actual city walls.

Further research is also needed to enhance tactile perception, particularly for small cavities that are difficult to discern through touch. Additional refinements, such as experiments with mesh displacement (Figure 12), should be explored. Recent studies suggest that specific model processing techniques can significantly enhance the tactile experience, making archaeological details more accessible to visually impaired users.

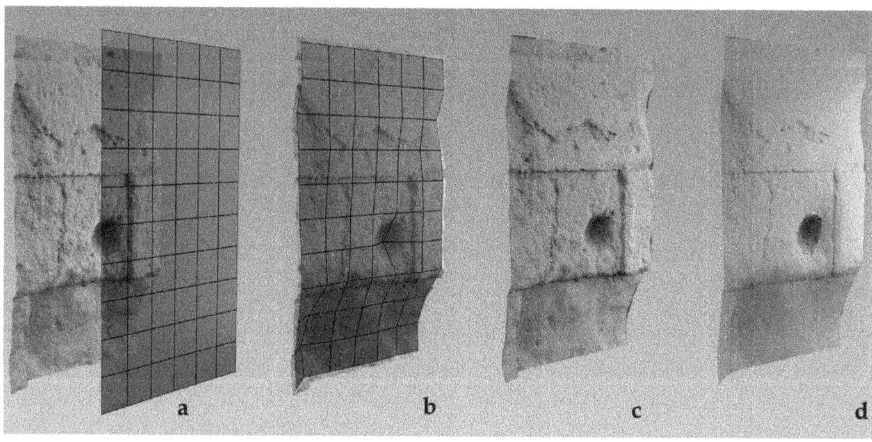

Figure 12. (**a**) High-poly model and low-poly quadrilateral mesh; (**b**) low-poly mesh constrained to the high-poly template; (**c**) displaced subdivision surface (SubD); (**d**) SubD with variable displacement intensity.

Tactile architectural models for blind and visually impaired users must exhibit specific physical representation features. In particular, they should be designed to emphasize and clarify the overall structure, while fine, high-frequency details should be either omitted or significantly subdued, as they may act as a form of "perceptual noise" that hinders rapid and accurate tactile perception. In the context of this case study, this issue necessitates the development of an ad hoc technique that enables the understanding of the general layout of the masonry and its individual blocks while simultaneously highlighting the presence of projectile-induced perforations.

The adopted technique is called displaced subdivision surfaces [29–31], which allows the use of 16-bit-per-channel OpenEXR displacement maps to accurately store the formal features of a high-detail model and transfer them onto a level-of-detail (LOD) surface, namely subdivision surfaces (or SubDs). The proposed solution involves applying the same OpenEXR displacement texture to the SubD model twice: the first application provides an overall level of detail to the tactile model without compromising the perception of its general structure (Figure 13). This texture is attenuated in intensity to prevent the 3D printing of excessively fine details that would be imperceptible to touch. The second application of the texture is used at full intensity but only in specific regions of the model—the areas corresponding to the impact deformations—thereby ensuring both legibility of the general form and tactility of the ballistic traces.

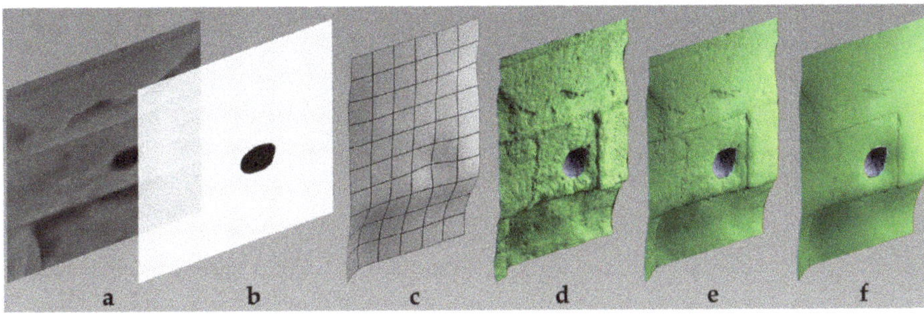

Figure 13. (**a**) Displacement map in OpenEXR format. (**b**) Mask generated by mapping ballistic imprints using Fiji software (version 1.54p); (**c**) Subdivision surface (SubD). (**d**) Segmented, displaced SubD model. (**e**,**f**) Differential displacement intensity: impact traces retain the original high-relief detail (purple), while background relief is reduced to enhance tactile perception (green).

5. Conclusions

In recent decades, 3D sensing technologies and digital modeling, including related advancements such as 3D printing, have become essential tools not only for CH documentation but also for enhancing cultural awareness and, importantly, for educational dissemination aimed at non-expert users. Virtual applications now play a crucial role in increasing both sensory and cognitive accessibility, providing practical and sustainable solutions. In many cases, these digital resources are complemented by physical prototypes, allowing a broader and more inclusive audience to engage with cultural heritage (Figure 14).

Figure 14. Rendering of a proposed informative museum exhibit of the northern walls, showing the analysis of the different types of existing cavities and stone balls and 3D printed casts of the cavities.

This approach is particularly valuable in cases where direct access is hindered, whether due to physical inaccessibility or the complexity of the subject matter, such as ballistic imprints, which are not easily recognizable, especially to the untrained eye. In these instances, virtual models and 3D-printed casts serve as powerful tools to aid understanding and interpretation.

Beyond analytical applications, the study demonstrated the successful adaptation of 3D models for museum purposes, resulting in both virtual reconstructions and physical replicas. The printed replicas were produced in both full-scale and reduced-scale versions, with some incorporating Braille descriptions to enhance accessibility for visually impaired users. This multi-sensory approach enables a wider audience to engage with archaeological evidence, overcoming physical and conceptual barriers typically associated with this specialized field of research.

These tactile and digital resources could also be integrated in situ, further contributing to the broader dissemination of archaeological knowledge. By offering direct and enhanced visualization of prototypes within their original context or through exhibition experiences enriched with popularized historical information about ancient weaponry, this approach facilitates deeper public engagement. Additionally, the possibility of evaluating the destructive effects of these weapons through an immersive virtual platform further enhances visitor interaction.

Author Contributions: Conceptualization, S.B.; methodology, S.B.; software, F.F.; validation, S.B.; formal analysis, S.B.; investigation, S.B.; resources, F.F.; data curation, S.B.; writing—original draft preparation, S.B.; writing—review and editing, F.F.; visualization, S.B.; Survey, mesh modeling and rendering S.B.; elaboration of figures: Figures 1–11 and 14 (S.B.) and Figures 12 and 13 (F.F.). All authors have read and agreed to the published version of the manuscript.

Funding: This research was supported by the project "SCORPiò-NIDI", CUP B53D23022100006 (DD n. 1012/2023), funded by the Italian Ministry of Research under the PRIN (call DD n. 104/2022) funding initiative.

Institutional Review Board Statement: Not applicable.

Informed Consent Statement: Not applicable.

Data Availability Statement: Research data are available upon request.

Acknowledgments: The authors would like to express their gratitude to the management and the appointed officials of the offices of the Pompeii Archaeological Park for granting authorizations for site access and survey operations.

Conflicts of Interest: The authors declare no conflicts of interest.

Abbreviations

The following abbreviations are used in this manuscript:

3D	Three-Dimensional
AC	Architectural Complex
AO	Architectural Object
AS	Architectural Section
CH	Cultural Heritage
LOD	Level Of Detail
SubD	Subdivision surfaces
SfM	Structure from Motion
SW	Siege Weapons
TLS	Terrestrial Laser Scanner
ToF	Time-of-Flight

References

1. *Las Casas Del Alma: Maquetas Arquitectónicas de La Antigüedad (5500 a.C./300 d.C.)*; Azara, P., Ed.; CCCB Centre de Cultura Contemporània de Barcelona: Barcelona, Spain, 1997; ISBN 84-7794-501-2.
2. Gavinelli, C. *Storie di Modelli Esibitivi e Critici: Modelli Storico-Critici di Rappresentazione Oggettuale di Visualizzazione Interpretativa*; Alinea: Florence, Italy, 1993; ISBN 2560005066993.
3. Sardo, N. *La Figurazione Plastica Dell'architettura. Modelli e Rappresentazione*; Kappa Edizioni: Rome, Italy, 2004; ISBN 978-88-7890-562-7.
4. Alberti, L.B. *L'arte di Costruire. Bollati Boringhieri*, 4th ed.; Giontella, V., Ed.; Bollati Boringhieri: Turin, Italy, 2010; ISBN 978-88-339-1882-2.
5. Scolari, M. *Il Disegno Obliquo. Una Storia Dell'antiprospettiva*; Marsilio Editori: Venice, Italy, 2005; ISBN 978-88-317-8617-1.
6. *Manual de Museología Interactiva*; Santacana i Mestre, J., Martín Piñol, C., Eds.; Ediciones Trea: Gijón, Spain, 2010; ISBN 978-84-9704-531-5.
7. Bruciati, A.; D'Alessandro, L.; Empler, T.; Fusinetti, A. VILLÆ (Tivoli, MiC). Percorsi di inclusione museale e accessibilità. In Proceedings of the DAI-Il Disegno per l'Accessibilità e l'Inclusione. Atti del II convegno DAI, Udine, Italy, 1–2 December 2023; Sdegno, A., Riavis, V., Eds.; PUBLICA: Alghero, Italy, 2023; pp. 508–521.
8. Bruciati, A.; D'Alessandro, L. Passepartout. Il museo di tutti per tutti. Un progetto di accessibilità delle VILLÆ. In *Aree Archeologiche e Accessibilità. Riflessioni ed Esperienze*; Anguissola, A., Tarantino, C., Eds.; Pisa University Press: Pisa, Italy, 2023; pp. 127–137, ISBN 978-88-3339-725-2.
9. Van Buren, A.W. Further Studies in Pompeian Archaeology. *Memoirs Am. Acad. Rome* **1925**, *5*, 103–113. [CrossRef]
10. Maiuri, A. *Studi e Ricerche Sulla Fortificazione di Pompei*; Monumenti Antichi pubblicati per cura della R. Accademia dei Lincei Hoepli: Milan, Italy, 1929; Volume 33.
11. Van Buren, A.W. Further Pompeian Studies. *Memoirs Am. Acad. Rome* **1932**, *10*, 7–54. [CrossRef]
12. Russo, F.; Russo, F. *89 a.C. Assedio a Pompei: La Dinamica e Le Tecnologie Belliche Della Conquista Sillana Di Pompei*; Edizioni Flavius: Pompeii, Italy, 2005; ISBN 88-88419-32-2.
13. Rossi, A. The Survey of the Ballistic Imprints for a Renewed Image of Unearthed Pompeii. *Nexus Netw. J.* **2024**, *26*, 307–324. [CrossRef]
14. Bertacchi, S.; Gonizzi Barsanti, S.; Rossi, A. Geometry of Wall Degradation: Measuring and Visualising Impact Craters in the Northern Walls of Pompeii. *SCIRES-IT-Sci. Res. Inf. Technol.* **2024**, *14*, 111–128. [CrossRef]
15. Rossi, A.; Gonizzi Barsanti, S.; Bertacchi, S. Use of Polybolos on the City Walls of Ancient Pompeii: Assessment on the Anthropic Cavities. *Nexus Netw. J.* **2024**. [CrossRef]
16. Lugli, A. *Museologia*; Jaka Book: Milan, Italy, 2009; ISBN 978-88-16-43033-4.
17. Negri, M.; Marini, G. *Le 100 Parole dei Musei*; Marsilio Editori: Venice, Italy, 2020; ISBN 978-88-297-0078-3.
18. Basso Peressut, L.; Caliari, P. *Architettura per l'Archeologia. Museografia e Allestimento*; Martinelli, C., Ed.; Prospettive Edizioni: Rome, Italy, 2014; ISBN 978-88-98563-06-7.
19. Rossi, A.; Gonizzi Barsanti, S.; Bertacchi, S. Naturali o antropiche? Misura e visualizzazione delle cavità murarie in cerchie urbiche/Natural or anthropic? Measurement and visualisation of wall cavities in city walls. In *Misura/Dismisura. Atti del 45° Convegno Internazionale dei Docenti delle Discipline della Rappresentazione/Measure/Out of Measure. Transitions; Proceedings of the 45th International Conference of Representation Disciplines Teachers, Padova, Italy, 12–14 September, 2024*; Bergamo, F., Calandriello, A., Ciammaichella, M., Friso, I., Gay, F., Liva, G., Monteleone, C., Eds.; FrancoAngeli: Milan, Italy, 2024; pp. 1957–1978.
20. Rossi, A.; Bertacchi, S.; Formicola, C.; Gonizzi Barsanti, S. Piccole indentazioni antropiche rinvenute nella riesumata cinta urbica di Cornelia Veneria Pompeianorum | The small anthropic traces found in the unearthed city walls of Cornelia Veneria Pompeianorum. *Disegnare Idee Immagin.* **2024**, *69*, 54–67. [CrossRef]
21. Rossi, A.; Formicola, C.; Gonizzi Barsanti, S. Ingegna Romana. Dalle fonti ai modelli, dai reperti alle ricostruzioni. *diségno* **2024**, 229–238. [CrossRef]
22. Busana, M.S.; Farroni Gallo, F. Aisthesis. Scoprire l'arte in Tutti i Sensi. 2021. Available online: https://www.museoomero.it/servizi/pubblicazioni/rivista-aisthesis-scoprire-larte-con-tutti-i-sensi/aisthesis-numero-17-settembre-2021/il-progetto-temart-metodologie-e-tecnologie-per-la-stampa-3d-di-riproduzioni-tattili-di-maria-stella-busana-e-francesca-farroni-gallo/ (accessed on 21 March 2025).
23. Leica Geosystems Cyclone [Software]. Available online: https://leica-geosystems.com/it-it/products/laser-scanners/software (accessed on 21 March 2025).
24. 3D Systems Geomagic Design X [Software]. Available online: https://www.3dsystems.com/software/geomagic-design-x (accessed on 27 February 2025).
25. Battini, C.; Fantini, F. Modelli tattili: Una questione di rappresentazione. In *Nuove Immagini di Monumenti Fiorentini, Rilievi con Tecnologia Laser Scanner 3D*; Bini, M., Battini, C., Eds.; Alinea: Florence, Italy, 2007; pp. 36–39, ISBN 978-88-6055-232-7.

26. Apollonio, F.I.; Zannoni, M.; Fantini, F.; Garagnani, S.; Barbieri, L. Accurate Visualization and Interaction of 3D Models Belonging to Museums' Collection: From the Acquisition to the Digital Kiosk. *J. Comput. Cult. Herit.* **2025**, *18*, 5:1–5:25. [CrossRef]
27. Spiess, F.; Waltenspül, R.; Schuldt, H. The Sketchfab 3D Creative Commons Collection (S3D3C) 2024. Available online: http://arxiv.org/abs/2407.17205 (accessed on 22 April 2025).
28. Balletti, C.; Ballarin, M.; Guerra, F. 3D Printing: State of the Art and Future Perspectives. *J. Cult. Herit.* **2017**, *26*, 172–182. [CrossRef]
29. Lee, A.; Moreton, H.; Hoppe, H. Displaced Subdivision Surfaces. In Proceedings of the 27th Annual Conference on Computer Graphics and Interactive Techniques, New Orleans, LA, USA, 23–28 July 2000; ACM Press/Addison-Wesley Publishing Co.: New York, NY, USA, 2000; pp. 85–94.
30. Catmull, E.; Clarck, J. Recursively Generated Bspline Surfaces on Arbitrary Topological Meshes. *Computer-Aided Design* **1978**, *10*, 350–355. [CrossRef]
31. Sheffer, A.; Praun, E.; Rose, K. Mesh Parameterization Methods and Their Applications. *CGV* **2007**, *2*, 105–171. [CrossRef]

engineering proceedings

MDPI

Proceeding Paper

Access to Digital Cultural Heritage: Exploring Future Perspectives Through Open Tools of Research †

Veronica Casadei and Giuseppe Di Modica *

DISI-Department of Computer Science and Engineering, University of Bologna, Viale Risorgimento 2, 40136 Bologna, Italy; veronica.casadei8@unibo.it
* Correspondence: giuseppe.dimodica@unibo.it

† Presented at the Conference "Discovering Pompeii: From Effects to Causes—From Surveying to the Reconstructions of Ballistae and Scorpiones", Aversa, Italy, 27 February 2025.

Abstract: In line with the research objectives of the SCORPiò-NIDI project, we aim to implement a software platform showcasing the digital models developed during the project. The goal is to develop dynamic and interactive user experiences, expanding access to cultural heritage through digital means, which become spaces for engaging and educational experiences. Using open-source frameworks, users can explore the complexity of Roman siege machines in an immersive way, interacting directly with the digital models. We will focus on the 3D model of the scorpion created by Dr. Claudio Formicola (University of Campania Luigi Vanvitelli), using the 3D modeling software Rhinoceros.

Keywords: digital cultural heritage; storytelling; 3D modeling; virtual tour; digital platform; interactive visualization; open source

1. Introduction

In 2014, the Council of the European Union [1] recognized digital resources as part of cultural heritage, alongside tangible and intangible assets. This represents a significant change in perception; digital heritage is now seen not just as replicas of physical originals but as independent entities that offer unique ways to access knowledge. This acknowledgment allows for innovative methods of experiencing and interpreting heritage, enhancing access for a broader and more diverse audience. Indeed, digital technologies enhance and amplify the ways cultural heritage can be accessed, shared, and appreciated, offering new opportunities for different communities to explore and engage with it. The Guidelines for the Digitization of Cultural Heritage [2] have also identified the enjoyment and enhancement of heritage assets as objectives of digitization, alongside the conservation of originals, scholarly study, and the recovery of previous digitization efforts.

However, for these digital cultural resources to acquire full meaning and become effective tools for cultural transmission, it is essential to integrate a narrative layer that contextualizes and brings them to life for users. The aim is not only to preserve the materiality of heritage but also to understand and convey the 'why' behind its existence and its value in collective memory.

In the context of the SCORPiò-NIDI project [3], the educational and scientific value lies in the unique opportunity, provided by the distinctive conservation history of Pompeii, to observe and communicate with absolute certainty the ballistic effects attributable to the artillery of the late Roman Republic, used by Lucius Cornelius Sulla during the siege of the city in 89 BC [4,5].

The section of the city walls between the Vesuvius gate and the Herculaneum gate still bears the impact marks caused by the projectiles launched by Sulla's artillery [6]. These imprints, preserved for nearly twenty centuries beneath volcanic ash, are indisputable evidence of the power achieved by Roman torsion artillery as early as the first century BC [6].

The project aims to enhance and disseminate these findings, providing the public with a dynamic reconstruction of historical events, such as the city's siege and the phases of projectile launching. It also seeks to facilitate the understanding of the technical-scientific evolution of Roman military artillery, offering an innovative perspective on Pompeii's history, traditionally known for its urban layout [4].

To ensure that this knowledge extends beyond academic circles, there is a pressing need to make these contents accessible and engaging for a wider audience of cultural heritage enthusiasts. Active engagement becomes a key element in fostering public involvement [7]. This means moving beyond passive reception towards a direct interaction that allows the public to explore, experiment, and interact with heritage narratives in immersive and dynamic ways. In this sense, digital technologies become more than mere communication tools; they enable the generation of new meanings related to heritage through the elaboration and reinterpretation of information.

This interactive approach materializes through digital storytelling, which transforms data, surveys, images, and 3D models into interactive experiences. Storytelling is the art of narrating stories, a human skill deeply rooted in oral traditions dating back to prehistoric times [8]. It has endured through the ages, transcending the simple transmission of information. Storytelling allows us to make the past accessible and meaningful in the present [9], offering a "story to tell" that guides visitors through the entire cultural experience. Thus, engagement becomes dynamic, immersive, and significant, turning the user from a passive spectator into an active participant, capable of transforming from a distracted passerby into an engaged observer [10].

Creating a narrative system that fully involves the user and makes them an integral part of the story means allowing them to participate in the creative process of digital heritage, fostering the co-creation of inclusive narrative forms. In this way, storytelling plays a key role in addressing contemporary challenges in cultural heritage, tackling crucial issues such as inclusion and equity [11].

In research, it is important to blend scientific rigor with storytelling. This approach allows cultural heritage to be presented as both an object of study and an engaging, interactive experience while upholding its authenticity and credibility. Narration is a blend of knowledge, technique, and art [10]; it is necessary to narrate while constantly engaging with scientific research. This paper is structured as follows: Section 2 provides an overview of commodity software tools used in the ATON project, specifically ATON [12] and Blender 4.3 [13]. We explore their functionalities and integration to meet project needs. Section 3 details the methodologies, completed activities, and future developments aimed at optimizing the valorization and accessibility of 3D models for an immersive user experience. We explain the utilization of these tools to achieve the project's objectives. This paper concludes in Section 4.

2. Materials and Methods

When storytelling is integrated into a digital ecosystem, content becomes dynamic, accessible, and customizable, adapting to different audiences and platforms. A digital ecosystem is a collection of tools, technologies, data, and processes that work together to create, manage, distribute, and consume digital content. The idea of an ecosystem implies interconnection and interoperability, with the common goal of disseminating cultural heritage. This is achieved not by a single software but through a network of tools and platforms [2].

The digital ecosystem developed for the dissemination of the SCORPiò-NIDI project mainly uses the ATON framework [12], Blender software 4.3 [13], and the virtual tour as an integrated tool. In a later phase of project implementation, the creation of a dedicated website is also planned, which will serve as a hub for multimedia and informational content. This integrated strategy aims to enhance accessibility, distribution, and engagement in the dissemination of scientific knowledge within the digital ecosystem.

2.1. ATON Framework

ATON is an open-source framework designed, developed, and coordinated by Bruno Fanini (VHLab, CNR ISPC ex ITABC) that enables the creation of cross-device Web3D/WebXR applications [12,14]. These applications are oriented towards online publication and interactive presentation of virtual scenes and objects for cultural heritage. They are accessible from a wide range of devices, from smartphones/tablets to laptops/PCs and immersive Virtual Reality tools (head-mounted devices) [15]. Users can access these applications without any third-party installation by simply typing in a URL from any platform [14].

One of the strengths of the framework is its scientific approach, which originates from the research sector and has evolved through numerous national and international research projects, aligning perfectly with the philosophy of our project [16,17]. Its open-source license makes it an ideal solution because it is accessible to a wide range of researchers, avoiding the costs associated with proprietary software and facilitating maintenance and updates. Moreover, the main rendering system, based on Three.js [18], supports the PBR (Physically Based Rendering) model, which is essential for presenting cultural heritage because it enables a realistic simulation of materials and how they react to environmental lighting conditions.

Another key element of the framework is the dynamism of the front-end and architecture, which allows hosting and distributing custom web applications [14]. The framework offers three usage options:

- Integrated front-end "as is" (if it meets the requirements) to present 3D models and scenes to end-users without the need for any code development;
- Custom extension, if one wishes to expand the front-end functionality;
- Develop and deploy a custom web app through the plug-and-play architecture, taking advantage of ATON's features and building a tailored user interface (UI) with the desired functionalities [12,14].

For the objectives of the SCORPiò-NIDI project, it was decided to integrate the basic software functionalities with the ability to develop a customized web app, thus expanding the ways of presenting 3D content, the CSS style, and the user interface. The official API documentation is also available on the framework's website, providing clear instructions for facilitating web app customization.

There are several scenarios in which ATON can be distributed, both in terms of hardware used and connection mode. Regarding hardware, ATON can be installed on a wide range of devices, from laptops for classroom experiments with a local network, to small servers in laboratories, and even to larger infrastructures with advanced hardware for complex installations [12,14].

From the network proximity perspective, the framework can be used in the following ways:

- Offline: ATON is installed directly on the device on which it is consumed, without requiring any connection, such as for use in a kiosk or touchscreen display in a museum;
- Local network: ATON can stream an interactive experience via an access point to nearby users, who can use a variety of devices without needing the internet, but with a configured local network;
- Online: via a remote server, where ATON can be hosted, allowing global access to content from anywhere in the world [12].

As a framework designed for the dissemination of cultural heritage, ATON offers several features. Among these is the ability to add semantic annotations by linking a subsection of a 3D object or scene to related information [19], which is presented to end-users in the form of texts, images, YouTube videos, audio, embedded pages, and more [12,14].

Other key elements of the framework include the separation and hierarchy between the concepts of collection and scene, which offer various advantages in terms of storage, reuse, and updating of loaded elements [12,14]. A collection is a set of elements added to the framework through a user profile, such as 3D models, panoramas, audio sources, etc., which can be used to create a presentation or an interactive 3D space [12,14]. The main format adopted for 3D models is glTF [20], which, due to its interoperability, provides smooth integration with various tools and 3D software engines (such as Blender, Maya, and Unreal Engine).

A scene is a configuration and integration of one or more elements from the collection, stored as a JSON file, and assigned a unique identifier, which can be used by any web app based on ATON to reference the specific 3D scene [14].

Thanks to this separation, the scorpion model can be added to the collection only once and subsequently reused in multiple scenes. For instance, in the context of enabling multilingual access, it would be possible to create two distinct scenes—one in Italian and one in English—without duplicating the model, but simply differentiating annotations, textual content, or interface elements specific to each language.

Other platforms that have overlapping concepts of a 3D model and a scene require complete reconstruction of the scene for each model update, which is time-consuming and resource-intensive. This limits flexibility in content management. However, since this is an ongoing research project, we can easily update the 3D model, including textures, details, or new components, by simply uploading the revised version into the collection. All scenes referencing the model will automatically reflect these changes without manual reconstruction [14].

2.2. Blender

Blender is a free and open-source program designed for creating three-dimensional content [13]. It supports the entire 3D production process, including modeling, rigging, animation, simulation, and rendering. Thanks to its versatility, the software allows for the detailed integration of visual and narrative elements, facilitating the creation of engaging and accurate virtual environments.

In the SCORPiò-NIDI project, the workflow in Blender involves the following:

- Importing the 3D model into the software. The model, created by Dr. Claudio Formicola using Rhinoceros 8.0, was previously exported in .glb format [20,21];
- Defining textures and materials to realistically recreate the visual and chromatic characteristics of the original object (wood, metal, etc.) as shown in Figure 1;
- Lighting the scene to create a realistic setting;
- Rendering and post-production to produce photorealistic images for use in ATON, project communication, and storytelling.

- Animation, simulating the dart throw and the movement of the mechanical components of the scorpion.

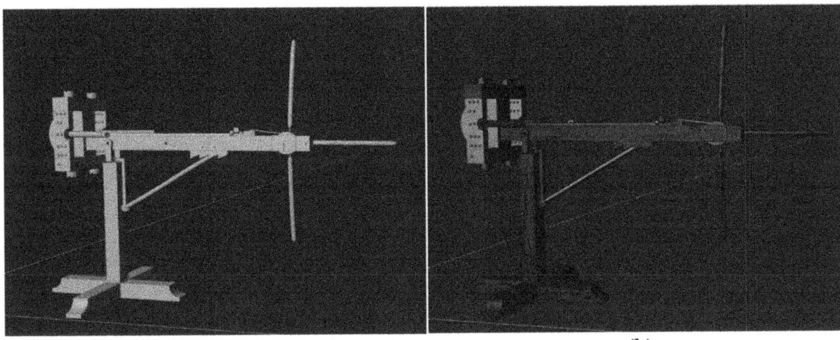

(a) (b)

Figure 1. The model pre- (**a**) and post-texturing (**b**).

In the context of cultural heritage, one of the crucial phases of creating 3D content with Blender is the definition of materials and textures [22]. These elements are essential for enhancing the authenticity of the models, optimizing the material properties, and realistically managing light interactions, to achieve a convincing and detailed visual appearance.

In Blender, a material defines the physical and optical characteristics of an object, controlling how it interacts with light and how it appears within a scene. The main properties of a material include reflectivity, transparency, roughness, and its composition. Textures, on the other hand, are 2D images applied to the surfaces of a 3D model, enriching its visual detail and simulating features like wood grain, stone irregularities, or reflection on a metal surface [22]. In this phase of the project, the focus is on optimizing the textures and materials of the scorpion's 3D model to accurately replicate the original materials, such as wood, metal and rope, as illustrated in Figure 2.

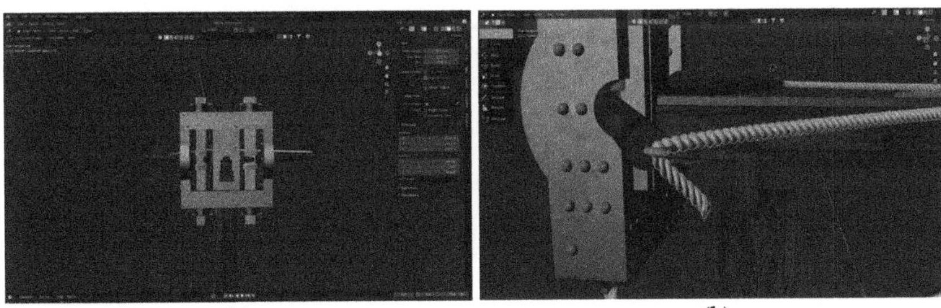

(a) (b)

Figure 2. Characterization of the model with threads (**a**) and ropes (**b**).

Blender offers a wide range of lighting tools, with four main types of lights: Point, Sun, Spot, and Area, each with adjustable parameters that allow for the creation of the desired lighting effect. Lighting is a crucial element in enhancing the realism of a 3D scene, as it plays a decisive role in the atmosphere and visual rendering of the project. Through careful management of lights, details can be emphasized, shadows can be played with, and lighting can be achieved that reflects physical reality, thus giving the model a sense of coherence [23].

Once the materials, textures, and lighting are defined, the next step is rendering, which is the process of converting the 3D scene into a 2D image [24]. Blender offers three main rendering engines, each with its own advantages and specific features: EEVEE,

Workbench, and Cycles. For the ongoing project, Cycles is being prioritized, as its path-tracing approach enables photorealistic results and effectively manages the interactions between light and objects [24]. Once the rendering is completed, Blender offers a powerful post-production phase through the compositor, which allows for further enhancement of the image with filters, visual effects, color corrections, and adjustments in contrast, brightness, and tones [24]. The final results can be exported in various formats (PNG, JPEG, TIFF, etc.), depending on the needs [24].

In addition to modeling and rendering, Blender was chosen for our project also for its advanced animation support. The ability to animate the 3D model of the scorpion, simulating complex movements such as the dart throw and interactions between the machine's various components, is one of the key features of the software. Blender supports a wide range of techniques for animating a 3D model, with the most common and straightforward method being the use of keyframes (Figure 3), which store the values of animated object parameters [25].

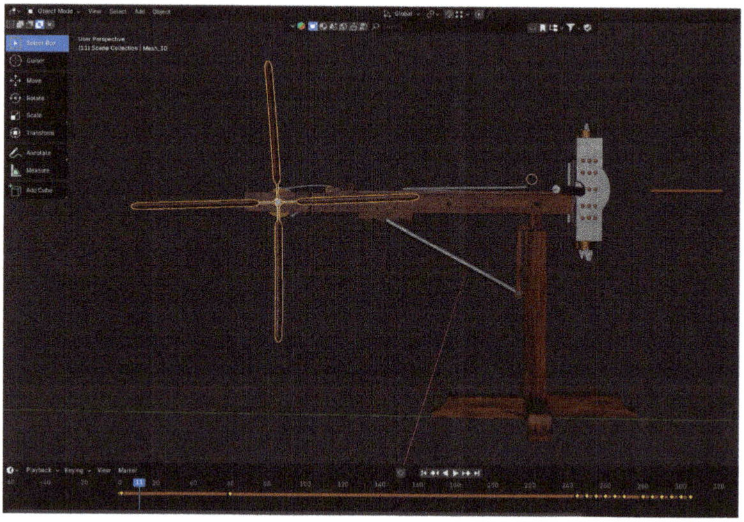

Figure 3. Static image of the dart throw created with keyframes.

The online Blender manual defines a keyframe as follows: "A Keyframe is simply a marker of time which stores the value of a property" [23].

A keyframe is therefore a specific point in time within an animation that records the value of a particular property of an object, such as its position, rotation, scale, or any other parameter that can be animated over time. When two or more keyframes are present, the software creates an intermediate animation by calculating the values between them to achieve smooth and realistic movement [25]. In conclusion, Blender offers a complete workflow with 3D modeling, material definition, lighting, rendering, and animation, making it ideal for the SCORPiò-NIDI project.

3. Results and Discussion

To create user pathways aimed at ensuring the maximum valorization and accessibility of 3D models, to provide users with an immersive and interactive experience, the tools and functionalities described above were employed. Specifically, the ATON framework will allow for a multi-level dissemination approach, ranging from the visualization of 3D models in a gallery to the future creation of a virtual tour [26,27] of the city walls between the Vesuvius gate and the Herculaneum gate.

The first step involved uploading the 3D model of the scorpion into a gallery within the framework, allowing future visualization and manipulation by users, who will be able to rotate the model 360 degrees and appreciate all its functional details. Users will also be able to view additional information and content through semantic annotations. Currently, semantic annotations have been added to the model, both in spherical and freeform shapes, consisting of points interactively placed on the queried surfaces. This allows the user's attention to be focused on specific parts of the model that merit further informational exploration to facilitate understanding. This is further enriched by adding images, videos, infographics, and audio to the annotations [14].

A second level involves the insertion of the animated model of the scorpion into ATON. The animation, which will simulate the dart launch, will be created using Blender software, where the 3D model of the scorpion was imported. Using the keyframe technique described above, all parts of the machine will be animated to allow a realistic simulation of the launch. The animation can also be enhanced with additional elements that help to contextualize it in time and space, such as the 3D model of the city walls, which can also be inserted and integrated into the software. The animated model will be exported from Blender in .glb format and then reintegrated into ATON, where it can be supplemented with additional elements, as described for the static model.

Another level involves the possibility of creating a customized web app accessible from any device, without requiring any installation by the end-users [14,17]. This app will focus on creating a virtual tour of the city walls. It can be enriched with 3D models of the machines, as well as text and annotations, with a focus on the ballistic impressions visible on the walls. The web app could also include an additional communication layer, namely temporal lensing [28], which allows users to interactively compare artifacts, objects, buildings, or large archaeological sites in their current state of preservation alongside their hypothetical reconstructions. This additional functionality could include not only the model of the current state of the walls but also a model showing their appearance around the 1st century BC.

This would help the user understand the dynamics of the siege and how the visible impact craters on the walls should be considered as missed shots. The actual targets of the launches were not the thick walls, but the temporary wooden shielding devices placed at the outer boundary of the walls, used by defenders as protection.

Furthermore, the use of Virtual Reality (VR) and Augmented Reality (AR) applications [29], supported by ATON, could be an exciting future development for the project.

There are therefore several perspectives being pursued that can be integrated and implemented together to achieve results capable of best valorizing the meanings and objectives underlying the project.

4. Conclusions

The promotion of the SCORPiò-NIDI project represents a significant example of how digital technologies can effectively contribute to the valorization and dissemination of cultural heritage. Through the adoption of innovative tools such as the ATON framework, Blender software, and the creation of an integrated digital ecosystem, dynamic, interactive, and accessible user pathways will be developed.

The integration of digital storytelling with rigorous scientific documentation enhances the narrative's authenticity and provides the audience with a meaningful educational experience. Additionally, utilizing open-source tools promotes sustainability through transparency, collaboration, and innovation. The ATON framework, when used online, allows researchers to collaboratively create scenes, share access easily among team members, and facilitates effective remote contributions.

Additionally, in the dissemination of the project, it will be essential to prioritize accessibility to ensure that everyone can use it effectively. This not only enhances inclusivity but also encourages broader adoption among diverse audiences [30].

To achieve this objective, we should refer to the Web Content Accessibility Guidelines (WCAG) [31], which provide a comprehensive set of recommendations for designing accessible web content for desktops, laptops, tablets, and mobile devices. During the platform's development, the following aspects must be considered:

- Text contrast: To ensure readability, high contrast between text and background is crucial. The WCAG recommends a contrast ratio of at least 4.5:1 for normal text and 3:1 for large text [31];
- Text size: The text size should meet the minimum recommendation of 16px for the main body of text [31];
- Text descriptions for non-text content: It is important to provide detailed text descriptions for all non-text content. For instance, comprehensive descriptions should accompany each phase of the 3D animations and the movements of the machines. This will enable users to understand the operations even when using a screen reader [31];
- Subtitles and Italian Sign Language (LIS): Subtitles should be included for videos and audio recordings, as well as content in Italian Sign Language (LIS) and other international sign languages [31];
- Simple and intuitive UI: A user-friendly and intuitive interface is essential for ensuring a positive experience for all users [31].

By following these guidelines, we can create an inclusive platform that is accessible to a wider audience.

To ensure the project's success, it is important to thoroughly study the target audience and review the user interface (UI) and user experience (UX) before the launch [32]. Future actions should include monitoring the project's impact and gathering user feedback to continually improve the experience and enhance overall accessibility.

Author Contributions: Conceptualization, V.C. and G.D.M.; methodology, V.C. and G.D.M.; software, V.C. and G.D.M.; validation, G.D.M.; formal analysis, V.C. and G.D.M.; investigation, V.C. and G.D.M.; resources, V.C. and G.D.M.; data curation, V.C. and G.D.M.; writing—original draft preparation, V.C.; writing—review and editing, V.C. and G.D.M.; visualization, V.C. and G.D.M.; supervision, G.D.M.; project administration, G.D.M.; funding acquisition, G.D.M. All authors have read and agreed to the published version of the manuscript.

Funding: This research was supported by the project "SCORPiò-NIDI", CUP J53D23012930006, funded by the Italian Ministry of Research under the PRIN funding initiative.

Institutional Review Board Statement: Not applicable.

Informed Consent Statement: Not applicable.

Data Availability Statement: No new data were created or analyzed in this study. Data sharing is not applicable to this article.

Conflicts of Interest: The authors declare no conflicts of interest.

References

1. Council of the European Union. Council Conclusions of 21 May 2014 on Cultural Heritage as a Strategic Resource for a Sustainable Europe (2014/C 183/08) 2014. Available online: https://eur-lex.europa.eu/legal-content/EN/TXT/?uri=planjo:20140523-025 (accessed on 21 March 2025).
2. Drafted by the Central Institute for the Digitalization of Cultural Heritage—Digital Library of the Italian Ministry of Culture within the National Plan for the Digitalization of Cultural Heritage 2022–2026. Available online: https://commission.europa.eu/projects/national-plan-digitalisation-cultural-heritage_en (accessed on 22 March 2025).

3. Bertacchi, S.; Gonizzi Barsanti, S.; Rossi, A. Geometry of Wall Degradation: Measuring and Visualising Impact Craters in the Northern Walls of Pompeii. *SCIRES-IT-Sci. Res. Inf. Technol.* **2024**, *14*, 111–128. [CrossRef]
4. Rossi, A.; Gonizzi Barsanti, S.; Bertacchi, S. Naturali o Antropiche? Misura e Visualizzazione Delle Cavità Murarie in Cerchie Urbiche | Natural or Anthropic? Measurement and Visualisation of Wall Cavities in City Walls. In *Proceedings of the Misura/Dismisura | Measure/Out of Measure. Ideare Conoscere Narrare | Devising Knowing Narrating*; FrancoAngeli s.r.l.: Milano, Italy, 2024; pp. 1957–1978. [CrossRef]
5. Rossi, A. The Survey of the Ballistic Imprints for a Renewed Image of Unearthed Pompeii. *Nexus Netw. J.* **2024**, *26*, 307–324. [CrossRef]
6. Russo, F.; Russo, F. *Tormenta. Venti Secoli di Artiglieria Meccanica*; Stato Maggiore Esercito: Roma, Italy, 2002.
7. Viola, F.; Idone Cassone, V. *L'arte del Coinvolgimento: Emozioni e Stimoli per Cambiare il Mondo*; Microscopi, Hoepli Editor: Milano, Italy, 2017; pp. 1–7.
8. Bonacini, E.; Marangon, G. Lo storytelling Digitale Partecipato come Strumento Didattico di Divulgazione Culturale. *Cuad. Filol. Ital.* **2021**, *28*, 405–425. [CrossRef]
9. Vaglio, M.G. Lo storytelling per i beni culturali: Il racconto. In *Racconti da Museo: Storytelling d'Autore per il Museo 4.0*; Dal Maso, C., Ed.; Edipuglia: Bari, Italy, 2018; pp. 27–51.
10. Costa, S.; Cordera, P.; Poulot, D. Storytelling? Una narrazione a più voci. In *Storytelling: Esperienze e Comunicazione del Cultural Heritage*; Costa, S., Cordera, P., Poulot, D., Eds.; Bononia University Press: Bologna, Italy, 2022; pp. 1–7.
11. Dal Maso, C. Storytelling: Perché. In *Racconti da Museo: Storytelling d'Autore per il Museo 4.0*; Dal Maso, C., Ed.; Edipuglia: Bari, Italy, 2018; pp. 11–25.
12. ATON. Available online: https://osiris.itabc.cnr.it/aton/ (accessed on 26 March 2025).
13. Blender. Available online: https://www.blender.org/ (accessed on 26 March 2025).
14. Fanini, B.; Ferdani, D.; Demetrescu, E.; Berto, S.; D'Annibale, E. ATON: An Open-Source Framework for Creating Immersive, Collaborative and Liquid Web-Apps for Cultural Heritage. *Appl. Sci.* **2021**, *11*, 11062. [CrossRef]
15. Gonizzi Barsanti, S.; Caruso, G.; Micoli, L.L.; Covarrubias Rodriguez, M.; Guidi, G. 3D Visualization of Cultural Heritage Artefacts with Virtual Reality Devices. *Int. Arch. Photogramm. Remote Sens. Spatial Inf. Sci.* **2015**, *XL-5/W7*, 165–172. [CrossRef]
16. Balzani, R.; Barzaghi, S.; Bitelli, G.; Bonifazi, F.; Bordignon, A.; Cipriani, L.; Colitti, S.; Collina, F.; Daquino, M.; Fabbri, F.; et al. Saving Temporary Exhibitions in Virtual Environments: The Digital Renaissance of Ulisse Aldrovandi—Acquisition and Digitisation of Cultural Heritage Objects. *Digit. Appl. Archaeol. Cult. Herit.* **2024**, *32*, e00309. [CrossRef]
17. Pietroni, E.; Menconero, S.; Botti, C.; Ghedini, F. e-Archeo: A Pilot National Project to Valorize Italian Archaeological Parks through Digital and Virtual Reality Technologies. *Appl. Syst. Innov.* **2023**, *6*, 38. [CrossRef]
18. Three.js. Available online: https://threejs.org/ (accessed on 31 March 2025).
19. Ponchio, F.; Callieri, M.; Dellepiane, V.; Scopigno, R. Effective Annotations Over 3D Models. *Comput. Graph. Forum* **2020**, *39*, 89–105. [CrossRef]
20. Khronos. Available online: https://www.khronos.org/gltf/ (accessed on 26 March 2025).
21. Rossi, A.; Formicola, C.; Gonizzi Barsanti, S. Ingegna Romana. Dalle Fonti ai Modelli, dai Reperti alle Ricostruzioni. *Diségno* **2024**, *14*, 229–238. [CrossRef]
22. Hosen, M.S.; Ahmmed, S.; Dekkati, S. Mastering 3D Modeling in Blender: From Novice to Pro. *ABC Res. Alert* **2019**, *7*, 169–180. [CrossRef]
23. Blender 4.3 Manual. Available online: https://docs.blender.org/manual/en/4.3/index.html (accessed on 2 April 2025).
24. Belec, A. *Photorealistic Materials and Textures in Blender Cycles: Create Impressive Production-Ready Projects Using One of the Most Powerful Rendering Engines*, 4th ed.; Packt Publishing Limited: Birmingham, UK, 2023; pp. 351–363.
25. Brubaker, S. *Realizing 3D Animation in Blender: Master the Fundamentals of 3D Animation in Blender, from Keyframing to Character Movement*; Packt Publishing Ltd.: Birmingham, UK, 2024; pp. 4–17.
26. Zambruno, S.; Vazzana, A.; Orlandi, M. Tecnologia, Beni Culturali e Turismo: I Tour Virtuali (Virtual Tours) come strumento per una corretta comunicazione dei Beni Culturali. *Stor. Futuro* **2014**, *34*, 1–4.
27. Meier, C.; Saorín, J.L.; Díaz Parrilla, S.; Bonnet De León, A.; Melián Díaz, D. User Experience of Virtual Heritage Tours with 360° Photos: A Study of the Chapel of Dolores in Icod de Los Vinos. *Heritage* **2024**, *7*, 2477–2490. [CrossRef]
28. Fanini, B.; Ferdani, D.; Demetrescu, E. Temporal Lensing: An Interactive and Scalable Technique for Web3D/WebXR Applications in Cultural Heritage. *Heritage* **2021**, *4*, 710–724. [CrossRef]
29. Fanini, B.; Pagano, A.; Pietroni, E.; Ferdani, D.; Demetrescu, E.; Palombini, A. Augmented Reality for Cultural Heritage. In *Springer Handbook of Augmented Reality*; Nee, A.Y.C., Ong, S.K., Eds.; Springer International Publishing: Cham, Switzerland, 2023; pp. 391–411. [CrossRef]
30. Rubano, V.; Vitali, F. Making accessibility accessible: Strategy and tools. In Proceedings of the 2021 IEEE 18th Annual Consumer Communications & Networking Conference (CCNC), Las Vegas, NV, USA, 9–12 January 2021; pp. 1–6. [CrossRef]

31. Web Content Accessibility Guidelines (WCAG). Available online: https://www.w3.org/TR/WCAG21/ (accessed on 2 May 2025).
32. User Experience Questionnaire. Available online: https://www.ueq-online.org/ (accessed on 2 May 2025).

MDPI AG
Grosspeteranlage 5
4052 Basel
Switzerland
Tel.: +41 61 683 77 34

Engineering Proceedings Editorial Office
E-mail: engproc@mdpi.com
www.mdpi.com/journal/engproc

www.ingramcontent.com/pod-product-compliance
Lightning Source LLC
Chambersburg PA
CBHW062033220926
51850CB00010B/2529
* 9 7 8 3 7 2 5 8 4 6 3 9 9 *